DAIRY GOATS

DIANA GREGORY

ARCO PUBLISHING COMPANY INC.
219 Park Avenue South, New York, N.Y. 10003

To My Kids

Claudia, Chris, and Lisa

Published by Arco Publishing Company, Inc.
219 Park Avenue South, New York, N.Y. 10003

Printed in the United States of America

Library of Congress Cataloging in Publication Data

Gregory, Diana.
  Dairy goats.

I. Title   Goats

SF383.G74   636.3′91′4   76-235
ISBN 0-668-03938-8 (Library Edition)
ISBN 0-668-03941-8 (Paper Edition)

CONTENTS

I	WHY KEEP A GOAT?	5
II	BUYING YOUR GOAT	9
III	VARIETIES OF MILK GOATS	16
IV	HOUSING	27
V	FEEDING	38
VI	GROOMING	47
VII	MILKING	56
VIII	COOKING WITH GOAT'S MILK	69
IX	BREEDING, PREGNANCY, AND BIRTH	80
X	RAISING THE KIDS	91
XI	ILLNESS	98
	APPENDIX	120
	INDEX	125

CHAPTER I

WHY KEEP A GOAT?

In a world that is filled with machine-made products and synthetically enriched, plastic wrapped foods, many people are returning to a simpler way of life. People who have spent years living in cities are now moving to the country, or at least to the suburbs, where they are discovering a new and healthy way of life in growing their own fruits and vegetables and possibly obtaining fresh eggs from a half-dozen or so hens. However, they often overlook one of the most widely used foods and one that is easy to obtain . . . milk!

Milk is the only protein which nature intended solely as a food. A charcoal-broiled hamburger may be delicious, but nature intended it to be locomotion for a cow, not a food. In today's economy, milk is one of the most costly of all foods, considering how much of it is consumed by the average person each day, both in drinking and cooking. However, if you live in the country or the suburbs, good, fresh-tasting milk can be obtained for only pennies per day. The answer is, of course, what this book is all about . . . the modern dairy goat.

"But," you say, "goats stink and the milk tastes awful." Just when was the last time you smelled a goat, or for that matter, tasted goat's milk? Have you ever smelled a pen filled with dairy cows? Ugh! What goats need is a better public relations department. To be sure, the buck, or male goat, does have a certain aroma, but that is only during the mating season, and you are certainly not going to want to own a male goat when

5

This young Alpine doe shows the intelligence and quality of a future champion.

what you want is milk.

Since the dairy goat is actually a small edition of the dairy cow and because it can be kept in a limited amount of space, the dairy goat is the perfect answer to overcoming the high cost of milk and dairy products, as well as providing milk that is free from additives and other foreign substances. An average dairy goat will give approximately five or six quarts of milk each day, or about 150 gallons per year, at about one-sixth the cost of the processed product.

Although the milk of the dairy goat is similar in taste, and for the most part cannot be differentiated from that of the cow, it is considerably different in composition, and is considered very therapeutic by many doctors because of its excellent qualities. Numerous people have been brought back to good health by drinking goat's milk, and many babies and young children who have been found allergic to cow's milk have thrived and grown strong and healthy on a diet of goat's milk. It is also interesting to note that all of the essential amino acids needed for building muscles, skin, and other tissues are found in milk, along with the calcium that appears to be lacking in the modern diet so overly abundant with phosphorus-rich foods such as meat, fish, and soft drinks (most of which contain phosphoric acid, which in turn produces bone loss).

Along with the milk, there is a yearly bonus in the form of two or three new kids (baby goats). The kids are usually sold soon after they are born and the proceeds from the sale go to pay for the cost of feeding the doe (mother goat) for the entire year.

Two Nubian does sunning themselves in their yard.

Dairy goats are not just a passing fad or ecological movement. They have been serving mankind for thousands of years. The New Stone Age people considered them such an important part of their community that they gave them first class accommodations in their caves. They were mentioned in the Bible and played a vital role in Moses' trek across the sands. The great leader Mahatma Gandhi never traveled anywhere without his two favorite dairy goats to supply him with daily milk.

Dairy goats were introduced to America early in her history, when the settlers of Virginia brought their milk goats with them from across the sea. However, the first purebred strains were not imported until many years later; the first said to have been brought into the United States in 1893. The majority of our registered breeds derive from importations made since 1900 which have developed into the five major breeds responsible for much of today's commercial and private dairy produce.

Furthermore, goats make enjoyable pets: they are intelligent and loving animals. It is no wonder that many of the younger generation, so interested in the future of the world and the ecological movement, are finding that ownership of a dairy goat is a rewarding endeavor.

If you can keep a dog, you can probably keep a goat. They are loyal, clean, loving, curious, and will even sleep at the foot of your bed should you so choose.

CHAPTER II

BUYING YOUR GOAT

Let us suppose that you have decided that you would like to own a dairy goat. What should you look for and where should you look? First, before you begin your search, there are a few things you should know before saying, "I'll take the pretty brown one with the white spots."

Most goats appear to be made up of a series of lumps and bumps, and, according to goat experts, the lumpier and bumpier the better. A goat that is in this angular condition is said to have "good dairy characteristics." The reason for this is that if the body is to produce milk it cannot be doing other things at the same time, such as adding fat. That is why dairy cows always have a bony look, while beef cattle are as sleek and fat as a good steak should be.

From front to back and side to side, your goat should have the general appearance of an oblong hat rack. You should look for prominent hip bones and thin thighs that blend smoothly into legs with plenty of good strong bone to carry your goat through many years of milk production. Her middle should be long and her ribs quite well sprung, giving her room enough to fill her stomach with hay and still have room for a youngster or two. Her ribs should be long and far apart enough so that you can place one finger down between them. This is termed "openness." She should have good width of chest and depth of body to provide adequate lung capacity.

The udder should be about the shape (and preferably the size) of a volley ball and must be well attached at the front as

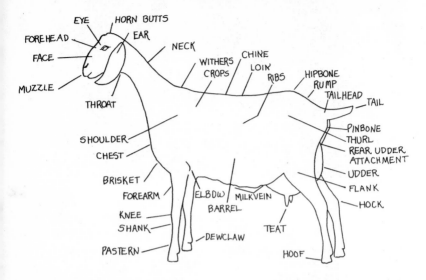

Anatomy of the dairy goat.

well as the rear. The bag, or udder, of the wild goat never grows very large as they only need to produce milk for a short period of time, just long enough to raise their youngsters. Since the dairy goat is tame and her main occupation is to give you milk, her udder will attain quite large proportions and will continue to grow in size each year, giving more and more milk until she reaches her peak of production around the fifth or sixth year. This is why it is of utmost importance for the udder to be attached to the body with strong, tight ligaments both in the front and the rear.

The amount of milk the individual goat will give depends somewhat on her particular breed, or combination of breeds. Some will give more than others, although those that give more produce a thinner milk than those that give smaller but creamier amounts. This is why breeders strive to maintain the purity of their herds and work to improve them with each new generation.

Purebreds are therefore a good buy as there is some assurance as to how much and what type of milk will be obtained, and there will always be a market for the youngsters

It would be difficult to beat the quality of udders, both in capacity and attachment, of the four does shown here. (Photograph by Bill Serpa; courtesy Laurelwood Acres.)

You can easily note the difference in udder capacity when comparing the younger doe in front to the older one in the rear; but each is a quality udder. (Photograph by Bill Serpa; courtesy Laurelwood Acres.)

that will be born each year. People are always anxious to find kids of good quality.

Purebreds are registered with an association such as the American Dairy Goat Association. They will have registration papers and will be tattooed in each ear with numbers and letters to correspond to those on their papers.

Occasionally a breeder will decide that he would like to try to improve his herd and will experiment with crossing different breeds. For example, he might hope to produce an offspring that would combine the doe's ability to produce rich-tasting milk along with the buck's ability to produce does that give greater quantities of milk. The resulting offspring is called a "grade goat." This youngster, of course, is not eligible to be registered as a purebred, but because both the parents are purebred, she can be registered as a "recorded grade." The price of these animals will not be as high as the purebred ones although they may produce a better grade of milk. They would therefore make an excellent choice for the backyard owner.

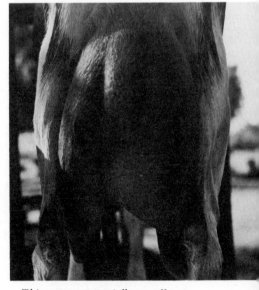

Although this udder shows both capacity and good rear attachment, it is a type to stay clear of due to the pendulous teats which make milking difficult.

This was a potentially excellent udder which was ruined for showing by poor handling. Nevertheless, the doe would still be an excellent buy for anyone interested only in milk production because of the capacity and well-shaped teats.

A goat that has no recorded history of her parentage is known merely as a "grade" and cannot be registered. Although she may give adequate amounts of good milk it will be somewhat difficult to sell her youngsters and in the long run she will, therefore, cost the owner more than if he had invested in a purebred goat.

When you begin to shop for a dairy goat, it appears that they are everywhere, populating almost every backyard, but for the most part these are not the type you are looking for. A goat is *not* something that you can order from a catalog, nor are there many listings for dairy goat farms in the Yellow Pages of the telephone book. Where then do you look?

One excellent way to learn more about goats is to attend a dairy goat show. There you can see the various breeds and listen to the judges make comments as to the conformation and various other points that go into making up the better goat versus the poorer one. There are usually several goats available for purchase at these affairs, and it is one way of talking to several breeders at once without having to travel a great distance. To find out when and where one of these shows will be held, you can check the list of clubs in the appendix (page 120) in this book for one near you and write or call for more information.

County and state fairs are also good places to meet goat owners. There you are more likely to find junior owners affiliated with 4-H, FFA, and other youth groups. They are usually more than enthusiastic about their animals and will happily talk to you as for long as you will stand still. Many of these owners will also have excess—yet quality—animals that they have brought to the fair for the sole purpose of selling. I picked up my April in this manner, paying $50 for a doe whose offspring now bring me up to $150 each. And she usually has at least three.

If neither of these suggestions fits into your plans, you might call the Farm Bureau or State Agricultural Office in your area and ask them to recommend a breeder. The people who work for these agencies are usually glad to help and will give you abundant information and whatever guidance you may require.

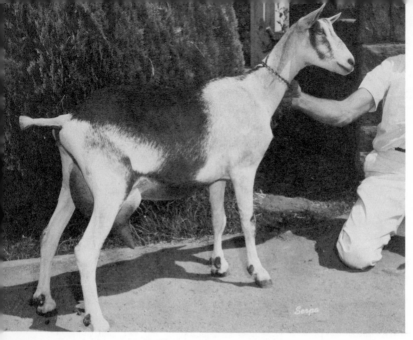

An excellent example of a grade *Alpine, demonstrating that quality can be found in the grade doe.*

Grand Champion K-Lou Wennie of Sunny Oak Farm. This seven-year-old Toggenburg doe is an excellent example of an aged doe with many productive years still ahead of her.

Whatever your choice, whether it be a young kid who will not be giving you milk for a year or two, or a matronly doe who is in the prime of her production years, choose carefully, for your goat will come to love you and consider you her family, and hopefully you will do the same.

CHAPTER III

VARIETIES OF MILK GOATS

Domestic goats come in two categories: (1) the Angora or mohair-bearing goat, and (2) the milk goat. Unless you are considering growing your own sweater, we shall dispense with the first class with a brief discussion.

Although there are over three million head of these strange looking creatures in the United States, few people know what they look like. The breed derives its name from Angora, a province in Turkey, in which the breed originated. Texas is the area in which most of these animals are currently concentrated, and if you know Texas, it's pretty easy to miss three million goats there.

Picture an Old English Sheepdog with horns and you've pretty much got the idea of what an Angora goat looks like.

Of course they do produce milk: the kids have got to get it from somewhere. However, this is not the breed you want. Now that we have covered the Angora goat industry, let us forge ahead into the intricacies of the dairy goat world.

There are five main breeds of dairy goats in the United States. They all produce milk and there is really very little difference in their production records, with the exception of that of the Nubian breed, which is a little lower, but which makes up for it in higher butterfat content.

The present breed record holders for a 305 day milk production test are: French Alpine 4826 pounds, Nubian 4392 pounds, Saanen 4905 pounds, and Toggenburg 5750 pounds. These figures fluctuate from year to year and, therefore, I feel it really doesn't pay to go by the latest figures off the press. How do you know that the goat you buy isn't a relative of the one that gave the most three years ago and is consequently producing like mad, not realizing that her breed is no longer the top contender?

Grand Champion Laurelwood Acres Yenda—National Champion Alpine in 1973 and 1974. (Photograph by Bill Serpa; courtesy Laurelwood Acres.)

ALPINES (FRENCH)

During the Napoleonic wars, the army had to carry much of its food with it, but the soldiers had to replenish their stores as they traveled. While passing through Switzerland they did a little midnight requisitioning in the way of goats, which they consequently took with them. It is thought that these hardy animals were the forebears of today's Alpines.

All of the registered French Alpines in the United States trace their ancestors back to three bucks and eighteen does that were imported by Charles DeLangle of California in 1922. If nothing else, they must be prolific.

The French Alpine is a large animal, rugged in build, and extremely hardy with few kidding problems. They have a classic head with a nose that any Greek god would envy. Their ears are erect and straight.

(Here I must digress as there is something I feel you should know. All of the breed registrations make a tremendous fuss about ears. As far as I can see or understand, ears have nothing to do with the function of milk production. In any case, if your Alpine has droopy ears she will most likely be banned from the goathood Blue Book.)

The range of colors Alpines come in is exciting. The descriptions of these colors add a definite touch of class to the breed.

There is Cou Blanc, which means she has a white neck; Cou Clair, which means a tan neck; and Cou Noir, which denotes a black neck. The body of the first two types is usually black or dark in color with the reverse being the case in the third. Sundgau describes a black coat with white markings. There is a Chamoisse which has a distinct resemblance to the wild chamois. They also come in other "flavors" such as cinnamon, strawberry, and vanilla.

The most important selling point of the French Alpine is their consistently high milk production. Some breeders also say that Alpines are the smartest of the breeds. I know that the first statement is fairly accurate, but as to the second, I think it all depends on what you consider "smart."

ALPINES (ROCK AND SWISS)

There are two other types of Alpines, although they are not as popular yet. Perhaps it is because they are not as numerous, or prolific, as the French variety.

The Rock Alpine is named after the breed's founder, Mrs. Mary Fairbanks Rock, who crossed the French Alpine with Toggenburg and Saanen stock and came up with a smaller, more delicate animal. They have multicolored coats with standard markings.

The Swiss Alpine is from Switzerland, which seems fairly obvious. The color is either chamois or a solid brown, ranging from light to deep red bay with black points.

The main difference between the French Alpine and the Swiss and the Rock seems to be the size, with the French being larger boned and rangier. However, all have good milk production records.

Grand Champion Laurelwood Acres Drusilla—Best Udder La-Mancha, 1973 National Show. (Photograph by Bill Serpa; courtesy Laurelwood Acres.)

LA MANCHAS

This is one of the lesser known breeds, but they seem to be gaining in popularity in certain areas. Their one outstanding characteristic is their ears—or apparent lack of them.

La Manchas were developed fairly recently in this country from a Spanish breed that was crossed with certain other purebreds. (They aren't telling which!) They come in a variety of color combinations. The main disqualification, according to the registry, is having anything larger than "gopher" ears. Well, some like them large and some like them small, but ears definitely are *the* thing in goats.

NUBIANS

Currently the Nubian is just about the most popular breed. Nevertheless, as with breeds of dogs, this may not be true by the time you read this book.

Theirs is a rather exotic background. They first originated in the upper Nile Valley, for which they were named. Legend has it that the first Nubian goats were a present from the King of Abyssinia to Napoleon III in the year 1860. They were sent from Africa to Europe to supply milk for a young hippopotamus. (What Napoleon III was doing with a hippopotamus in the first place is one of those things that tend to boggle one's mind. I am not enlightened on the reason, but it is a good piece of information for trivia collectors.)

Because the goats did not thrive in the damp climate, they were crossed with the native stock. It is from this crossbred foundation that the Nubians of the United States are descended, with the first ones immigrating about 1896.

The Nubian is a large, distinguished looking animal with long drooping ears and a roman nose. Some are born with horns and some without. The does fortunately do not come equipped with beards. They have a short, shiny coat that comes in a wide range of colors, the most common being black, tan, and red. They can be any of these colors, with or without white markings. Some have big white spots all over and appear to be dressed for a costume ball.

Grand Champion Laurelwood Acres Zayette—National Champion Nubian in 1973. (Photograph by Bill Serpa; courtesy Laurelwood Acres.)

Nubians are the most gentle of the breeds, but they also can be the most stubborn and can sometimes act like a spoiled child having a tantrum. They cannot stand to be left alone. If there are two of them living together and one of them is out of sight of the other, she will bellow until the other heaves back into view again. Incidentally, the bellow of a Nubian goat is very, very nosiy.

As stated somewhere previously, Nubians are not known for their large milk production, but the milk is so rich that it almost seems as though you are drinking half-and-half. Among Nubian lovers they are known as the "Jersey" of milk goats. So, if yours is a small family with a liking for rich milk, consider this breed.

However, if you don't plan to get two goats, prepare to spend a lot of time with the one you do get. Nubians need and demand a lot of love and attention.

Grand Champion Laurelwood Acres Blanca II—National Saanen Champion in 1973. (Photograph by Bill Serpa; courtesy Laurelwood Acres.)

SAANEN

The Saanen derives its name from the Saanen Valley in Switzerland, its native home. Ten Saanen goats were imported into this country by the United States Department of Agriculture in 1898.

Saanens have large strong bodies with udders to match. They are called the "Holstein" of dairy goats. They give large quantities of milk but it is not rich in butterfat. (They are at the other end of the scale from the Nubian in this respect.) Therefore, if you are always dieting, or simply have a large family to feed, this is the breed you should aim for. I understand that they are somewhat difficult to find at the moment as they are in great demand for commercial dairy use, but this may vary in different parts of the country.

Due to their all-white coloring, the Saanens are one of the prettiest of the breeds. What more could you want for a lawn decoration?

TOGGENBURGS

The "Togs"—as they are known in goat circles—take their name from the Toggenburg Valley in Switzerland, where they originated. Four animals of this breed were imported into Ohio in 1893, making them the first breed to reach our shores. Five years later the United States Department of Agriculture brought in sixteen more.

Although they are a smaller breed, with the average doe weighing in at about 100 pounds, they are hardy. Their ears are erect and their noses straight or somewhat dished. They are usually hornless.

The cost of a purebred doe, of any breed, currently runs anywhere from $150 to $300 and up, depending on breeding, age, and milk production.

Grand Champion Laurelwood Acres Bernadine—Toggenburg doe. (Photograph by Bill Serpa; courtesy Laurelwood Acres.)

RECORDED GRADES

If you cannot afford a purebred goat, consider the recorded grade. This is a doe with one registered purebred parent, either sire or dam, and the other of unknown heritage.

Technically, a doe can become a half-grade even if she has no registered purebred parent, but she does have to have either an "official Advanced Register production record," or a "Star Milker certificate," or to have been designated a "Permanent Champion," provided she conforms to the breed type and the specifications of that particular breed.

The first two items mentioned have to do with the amount of milk this doe has produced in one period of testing. This will probably send her way out of the class of the backyard goat anyway; at least in price.

If both doe and sire are purebred, with papers, but are of different breeds, the kids from such a mating will be classified as crossbreds and will be recorded as "experimental." What this means is that they will still be listed as Recorded Grades, but with a little more class. The registration papers are fancier, too, with a pretty little border. And, if it makes you happy, you can think of yourself as a part of an experimental program.

Recorded does can be "graded up" through successive breedings to purebred sires of one breed. The third such generation of continuous recorded ancestry will be classified as seven/eighths purebred and will then be allowed to register as purebreds of that particular type.

An "un-recorded grade" is one of unknown or unrecorded family but with definite breed characteristics. It is possible to have such a doe graded up. Whether it is worth the expense and time involved is something each individual must consider for himself.

A good recorded grade doe milker will cost from $85 to $150, depending upon age, quality, and milk production.

SCRUB GOATS

This is the bargain hunter's basement, and there really are some bargains here. All dairy goats give milk. Perhaps some of

PUREBRED FRENCH ALPINE

| THE GOAT named | MISCHIEF ACRES VALENTINE | | No. A212222 | Vol. 213 |

			Sire's Sire	Play Fair Sir Winston	A179373
Sire	Sal's Gals Guy Elm	A191385			
			Sire's Dam	Maefair Valentine of Fair Oaks *M (AR20)	A170169
			Dam's Sire	+*B Laurelwood Acres Yahtzie (AR19)	A155677
Dam	Jim's April	A167340			
			Dam's Dam	Jo's Sue	A149183

Description	Cou clair		
Sex	Doe	Date of Birth	February 14, 1975
Horn Information	Dehorned	Tattoo	RE:MA LE:I2

Bred By   Lisa Gregory, Placerville, California

Owned By   Mischief Acres, Placerville, California - March 14, 1975

Owned By   Christina Renee Marco, Placerville, California - April 25, 1975

MISCHIEF ACRES VALENTINE   A212222   Has been accepted for registry   April 9, 1975

AMERICAN DAIRY GOAT ASSOCIATION

Under the rules of the Association
Alterations to this certificate except as made by
the ADGA office, render it NULL AND VOID

Lyon Wilson Secretary

PRODUCTION RECORDS

Registration form from the American Dairy Goat Association.

them in this category are a little strange looking, but almost everyone except a goat fancier thinks goats are funny looking anyway.

A scrub goat is equal to a mutt in canine terms. There are usually very few clues as to parental heritage. The main disadvantage in purchasing a scrub goat is that you will not make any money off the sale of the kids, and since your goat does have to produce kids every year in order to keep the milk supply flowing, you might as well receive that extra few hundred dollars you can count on as the price for a couple of registered kids. Also, unless you really know what to look for in dairy goat characteristics, you may end up with a goat that will cost you more in the long run what with kidding troubles or a breakdown in her udder.

Nevertheless, it is entirely possible to find a goat that will produce record amounts of milk. Thus, if milk is mainly what you are seeking, making money off the kids is negligible, and if you wish to spend a minimum amount of money, the scrub goat may be just right for you.

A good scrub doe that is milking can be purchased for as little as $25.

Just one further bit of advice. If you like to know what you are getting, a purebred is of course your best bet. Not only will you have a good idea of what the milk production is now but also what you can expect to get in the future, and she will more than pay for herself on her first freshening. However, if you can't quite swing a purebred, the next best thing is a recorded grade. This could be an excellent choice since it is a chance to get the best of two breeds rolled up into one great little package.

Just remember, there are a lot of choices to make, and a goat is not just a goat, it is an investment that is going to pay heavy dividends and be an important member of the family as well.

CHAPTER IV

HOUSING

A goat is not a fussy creature. She will happily spend her days in a packing crate if it is comfortable. Her needs are quite basic. She requires a structure that is draft-free, warm, and cozy in the winter; and with adequate ventilation so that it is not too hot in the summer. In addition, she will need a pen to do her romping in, a few toys to help her pass the time, and something to put feed and water into. All of this can be accomplished quite easily, even if you have never raised hammer to nail.

To begin with, let us consider the basic structure—her house. If you intend to be a one-goat family, you might consider purchasing one of the larger, ready-made dog houses that are available through lumber yards and pet stores. If you tell the man that you are buying it for a large Great Dane or St. Bernard, you will avoid those funny stares and needless explanations. Once home, the only alteration needed is the addition of something to break the wind (remember, no drafts) in the doorway. You might consider hanging a burlap sack from the inside so that it falls down into the doorway when it is chilly, and still allows air to circulate freely. (You can remove it during the hotter months.) If you place the house facing away from the prevailing winds, and have a natural windbreak or shelter of trees, you probably would not even need the addition of this semi-door.

Another ready-made or quickly put together structure is similar to the dog house, yet allows you to go all the way with your imagination. This is a playhouse which usually comes in

A simple structure made from plywood that adequately shelters three does, made by their eleven-year-old owner.

sections and which almost anyone can put up in record time. You can decorate it to your heart's desire by painting Tyrolean flowers and cupids all over it. Just be sure to use lead free paint when doing your decorating as your goat will no doubt love your art work as well.

On the other end of the scale from the ready-mades are sturdy packing crates. I have a friend who whipped up quite a suitable domicile out of a piano crate. Although I don't know just how often you are likely to run across one of these, I'm sure you get the idea.

If you are handy with tools, you probably already have in mind a modified A-frame, complete with decking and louvered windows. I'll say no more. However, do keep in mind that you should build the entrance so that it is facing away from the winds.

For the flooring of your goat's little home, there are several alternatives. My preference is for plain earth. It is porous and will allow moisture to drain, leaving a fairly dry bedding behind. If you are both energetic and/or have a hardpan type of

A more elaborate structure which can house up to twelve does, but which was still relatively inexpensive.

soil, you might dig up about a foot of earth, put in a layer of gravel, and then replace the soil. Drainage should then be just about perfect. My second choice is wood flooring. (You will automatically have this if you have purchased a dog house.) This flooring has one distinct drawback as the wood tends to absorb odors and has to be constantly cleaned. Nevertheless, it is not as hard on the goat as the third choice, which is cement or concrete. It would not be a wise choice to deliberately make a concrete floor, but you may be considering utilizing an old shed or part of the garage and, therefore, the choice has already been made for you. Concrete is hard on the goat's udder and may cause bruises; is not porous and is consequently hard to keep clean; and it is cold in winter. However, all this can be overcome with a good heavy layer of bedding.

The type of bedding you select is pretty much determined by what is readily available in your area for the least amount of money, and whether or not you plan to employ the used bedding as fertilizer for your organic garden.

The bedding materials prevalent in my own area are wood

A kid house, used for kids up to six months of age. A large dog house would also serve just as well.

shavings and straw. On the surface, wood shavings would appear to be an ideal material as they are cheap, absorbent, and will work nicely into a compost. However, goats have a tendency to eat their bedding. I did not realize this for some time, but eventually it dawned on me that the bedding kept disappearing, and there seemed to be no logical place for it to go other than the goat's stomach. It probably does little harm to the goat, but I have visions of splinters everywhere; thus straw made its debut in our barn.

I have found straw to be quite suitable as it will not hurt the goat if she eats it, is fairly cheap, and mulches down well in a compost heap. Also, it is quite absorbent, thus keeping the quarters fairly clean and dry. I have found that I use about a bale of straw every six to eight weeks at a current local cost of about $2.00 a bale.

If you wanted to pick the best bedding, it would most likely be peat moss. It is absorbent, clean smelling, and an excellent additive to any garden. However, the main drawback is its high cost. It also takes about ten sacks to do the job of one bale of straw.

If you live in the South, peanut shells are quite good. I cannot speak from personal experience as to their excellent bedding qualities, but they are recommended by the U.S. Department of Agriculture.

If your flooring is dirt, you can build up a rather thick bedding of any material by removing the top layer and adding a lighter layer of new materials on top. Any moisture usually goes through the bedding and into the ground beneath. This added layer build-up makes for a warm and comfortable bed and helps keep the goat's house free of drafts and ground dampness.

I am sure that there are endless possibilities for bedding materials, but these four choices will serve to start your thinking processs. It depends on whether you live near a sawmill, peanut butter factory, or bog. Just get something down for her to lie on. It will make your job much easier and her life a lot sweeter.

BUILDING A PEN

No goat wants to be fenced in, but you can't just have her wandering the streets: the neighbors will object.

Again, how are you at putting up structural things? Are you going to dash out to the nearest lumber yard and order the first dog house you see, or did you immediately begin drawing up plans in your mind when confronted with the knowledge that your goat would need shelter? If you are of the latter bent, a beautiful white post and rail fence would be perfect, and you will have a complex that will be the talk of the neighborhood. Just be sure that the fence is at least four feet high (remember that goats are agile little creatures used to jumping about the Alps), and that the boards are close enough together so that your goat will not sneak out between them. Use only edible paint: aluminum is a good choice, or whitewash.

If, on the other hand, you are on the other end of the handyman scale, your complex may still be the talk of the neighborhood, even to the point of a petition or two. Do not let it worry you; your neighbors will soon grow used to its presence.

For you, the best bet is wire fencing. This comes in rolls of several hundred feet or more. There are two basic types, the woven type and the mesh type. The mesh type is usually a more closely spaced fencing and is a little stronger, and hence will last a little longer. You might note that the heavier the gauge, or thickness of the wire, the lower the number. Thus, 9 or 11 gauge wire is a lot stronger than a 12½ or 14½ gauge wire. The 47 inch height would be best for your purpose.

To hold up this roll of fencing you will need some sort of posts. Your feed dealer will also carry metal posts, along with the fencing. They come in various heights and have little attachments that the wire fits onto. Be sure to allow for enough length to put at least 1½ to 2 feet into the ground with some left over on top so that the fence won't slop over. A post that is about six feet would be just about right. If you wish to spend a bit more money and obtain wooden posts, your lumber dealer should have some 4 inch x 4 inch redwood posts. Again, aim for six foot lengths, and don't forget to treat the ends with creosote before you plant them in the ground.

Your goat should have a minimum of 200 square feet to play in, and more would make her ecstatic.

LOCKING THE BARN DOOR

One thing you'll soon find out is just how intelligent your goat can be when it comes to wanting to roam, and that is one thing they love to do. To gain her freedom, she will quickly learn to untie almost any knot you can devise, unlatch any latch you can buy, and burglarize any lock that can be picked with the teeth. It is therefore wise to place your lock on the outside of the gate where she cannot reach it with her teeth.

When you think you have her safely tucked away, find a hiding spot and observe her for a period of time. When I had settled April in her new home, I thought all was tight as a drum, but I kept finding her out in front of the house watching the traffic go by or stripping the leaves off the peach tree.

Well, I couldn't figure out how or where she was getting out. There were no holes in the fence and it was snug to the ground, and the gate was latched. This time when I put her back, I took cover and watched for her to make her escape. For a while she lay in the sun, supposedly chewing her cud, until she thought the coast was clear. She then casually rose and strolled into her little house. Not a minute later I saw one of the boards rise up, and out she walked as nice as you please.

How clever can you be! She had found a loose nail, worked it out, and there she was with her own secret exit. I applauded her form, but proceeded to put a baseboard around the outside wall of her house, while she went back to sitting in the sun and dreaming up new ways to escape.

HER PLAYGROUND

Remember that goats like to climb mountains and rocks and things Alpine in nature, and if given a chance, your goat will spend many happy hours on top of your car or house. You should, therefore, provide her with substitute structures. These are easy. Any old, solid, wooden crates will make a mountain that she can climb up and down for hours at a time. Just bolt them together (no nails) and make sure they are not near the fence so that she can leap to feedom.

If you are lucky enough to be able to obtain one of those round wooden cable reels from the telephone company, this will make a great mountain plateau to jump on and off. It will also provide a place for her to lie in the sun, or underneath in the shade if that is the desire of the moment.

A teeter-totter can be made with an old sawhorse or nail keg for a base and a sturdy 2 inch x 8 inch plank. An empty nail keg is also a lot of fun to roll and butt around.

Hopefully, the place you have chosen to put her home is one in which there are one or more shade trees. You can use the limbs of these to hang an old tether ball from, or an empty bleach bottle that has been partially filled with loose pebbles so that it acts like a rattle.

Another idea is a kind of fun snack. When cutting limbs from small trees or pruning your bushes, take the larger, tree-like limbs and "plant" them in your goat's yard. She will dote on you for these treats. Just be sure that it is a plant that is edible. Don't forget to plant your Christmas tree in her yard next December 30th, either.

These little treats are great, but you'll have to do better than that in the feeding department.

A FEEDER IS HER FAVORITE TOY

Even those with less dexterity than most can manage to put together something that any self-respecting goat will not quibble over. What is desired is a container that will hold enough grain and hay to make her happy, will keep it from coming in contact with the ground (she will then dismiss the food as inedible), and will keep out such elements as rain, snow, and bird droppings. An easy-to-assemble feeder can be made after a quick visit to your neighborhood junk shop or nearest garage sale. What you need is a small wooden table. Turn this upside down so that the underside of the tabletop becomes the floor of the feeder, and add some pieces of 1 inch x 4 inch wood to keep the grain and hay from spilling out. Then, to keep out the elements, you can nail a piece of plywood to the legs, which now act as supports for the roof. The total time required to build my own version was about 25 minutes and the cost was under $2.50.

Goats will consume from one to three gallons of water a day, so be sure to obtain a container that will hold at least that much. She should have clean water each day, so anything bigger will just let you get away with not changing the water often enough. She wants a drink, not a bath. I found my goat's water pan by haunting the Salvation Army store. It is a perfectly good, old fashioned enameled dishpan and cost a mere 35 cents.

As far as a place for the salt block is concerned, anywhere will do as long as it stays dry; they have a tendency to melt in the rain. You will probably be buying a small five pound block. Some feed stores have brackets for this size that you can attach to the wall of her house or to a nearby fence post. I simply toss mine in the feeder where it seems to do just fine.

This combination feeder and pleasant resting spot was put together using scrap 2 × 4's, 2 × 6's, and plywood. It is sturdy, will prevent feed loss and spoilage, and only took a few hours to make.

GRAZING

If you are the average crabgrass owner, you will undoubtedly want to let your goat "mow" your lawn occasionally. This is excellent, both for your goat and for you. She will make a nice even path through your lawn and her milk and digestive system will be all the better for the fresh grass. There are, however, some important rules to follow.

Do not tether your goat out and go off and leave her. There are two accidents that can happen very quickly, and both are lethal! You should always be within hearing distance of a bleat for help.

Dogs are a menace to goats. They are natural enemies. A goat will infuriate a dog by butting him, and the dog will retaliate in the only way he knows, with the sad result that the goat can do nothing to save herself. The goat will usually be so injured that she will either be dead by the time you return, or will be so far gone that she will probably have to be destroyed. This can happen with the most lovable dog in the neighborhood, or even your own, if so provoked.

The second menace is the goat herself. If she gets her head caught in something, a goat will fight until she is free, even if her head is left behind. If she becomes ensnared in a rope or tether, she will struggle until she has strangled herself.

If the lawn is what you want to cut, sit in a nice sunny spot and let her go to work; or be within earshot in the house, but do keep your ears open. If you love her, and I hope you will, treat her as you would any two-year-old that you have put out to play in the yard for a while.

If you have plenty of land and are considering taking her grazing one step further, you might consider planting a small permanent pasture. Alfalfa, clover, soybean, and lespedeza are the most frequently used. For the best pasture grasses for your particular area, you might consult your County Agent, Farm Bureau, or State Agricultural Experimental Station. They are always anxious to please, and all information is freely given.

If you have trees in your pasture, or on your lawn, that you wish to protect from the eager appetite of your goat, there are a few devices which will help. One way is to wrap chicken wire around the trunk. Since this is not too aesthetic, there are two other methods to consider. Painting the trunks regularly with asphalt roofing cement is one. Another is to paint the trunks with a manure tea. Take a large pail, fill it half way with goat manure, add water to the top, and leave the mixture to age overnight. In the morning mix it well and apply two coats to the tree trunks and limbs with a paint brush. You will not notice anything, but your goat definitely will and will steer clear of that tree.

THE END PRODUCT

This is a good time to mention what wonderful things manure tea can do for other growing things. If you are a gardener, and you probably are if you have decided to get a goat, use this tea to pour around your plants. Be careful, as it is very powerful. Carefully pour it around the roots as it will probably burn the leaves if you get it on them. Stand back and watch your plants leap up wildly.

CHAPTER V

FEEDING

Each time I have picked up a text on animals and flipped to the section on feeds I have been confronted with an enormous fund of knowledge on TDN's (Total Digestible Nutrients), charts on gross abnormalities due to vitamin and mineral shortage, and tables showing how to mix quantities of grain rations that would feed a good sized dairy herd. Unless you are studying for a degree in animal husbandry, I feel that it is a waste of your energy to attempt to digest this overabundance of nutritional knowledge.

What you really want to know is what to pick up at the feed store. It is basically very simple. Your goat will eat hay, grain, a dash of salt, and a few potato peelings or leftover fruit salad.

What this chapter contains, therefore, is a smattering of knowledge to take with you on your first trip to the feed store. After that you're on your own. After all, do you consult a nutritional specialist every time you run to the local supermarket to pick up a tidbit or two for your dinner? And, as for those nutritional specialists, you will find them hand in glove with the commercial feed companies where they are paid to mix up those feeds you will be buying.

A GOAT IS SOMEWHAT DIFFERENT

I think a short internal tour of your goat would help to clarify why she consumes the type of feed she does. What follows is brief and about as complicated as this chapter is going to get.

Your goat is what is known as a ruminant. A ruminant is a cud-chewing animal having a four-compartmented stomach capable of digesting roughage such as pasture, peach trees, and poison ivy. A ruminant can use up to 50 to 80 percent of this roughage in her diet. The other less fortunate simple-stomached animals such as the horse must eat a smaller amount of these bulky feeds and a larger amount of concentrated feeds and grains.

The rumen, or fermentation vat, represents nearly 80 percent of the total stomach size. This portion of the stomach is inhabited by countless bacteria and protozoa which supply the enzymes that break down the fiber and other parts of the feed in order to build protein from the simple nitrogen compounds and manufacture many of the vitamins needed by your goat. In contrast, the simple-stomached animals must have most of their vitamins supplied to them.

However, these bacteria must be properly fed to do their job, and this is where you come in.

HAY

Unlike other classes of livestock, your goat will thrive beautifully on brushy hillsides and weed covered lots. But, once she has cleared your property, you must consider the purchase of commercially grown hay. The kind you buy will usually be decided for you by the area you live in. You can safely feed your goat any good quality legume hay such as alfalfa, clover, soybean, and trefoil. (A legume, by the way, is simply a grass that has the added attraction of a little more protein, in that it also produces a seed such as a pea or bean.)

A mixed hay containing legumes and grasses such as timothy, red top, sudan, and brome grasses will also do nicely, if it contains at least 50 percent legumes. If only grass hay is fed, you will have to increase the amount of grain to balance the protein ration. It is therefore probably more economical to stick to the legumes, if possible.

How to Determine Hay Quality

If you live in a town where there is only one feed store, you're set in where you must purchase your feeds, but if there are two, you should make an attempt to determine which provides the best rations.

In purchasing hay, cost and quality factors will vary widely, but a bleached out, year old bale is no bargain at any price.

Try to examine the hay before you buy it. A reputable feed dealer will be more than happy to discuss the quality with you. Occasionally, though, even the best feed dealer will have a bad batch. Several factors may be involved, such as a bad year, weather-wise, or sudden shortages. In that case you might be better off buying a pelleted feed, but more about that later.

Coarse-stemmed hay will be mainly wasted, as your goat will greedily push the stems into the floor in her search for the more tender and tasty leaves. Once it is on the floor, she will discard the hay as being unfit to eat. Your first thought should be to look for leafiness and for lots of good green color.

Leafiness is important, not only for its taste appeal, but also for the high protein and vitamin content found in the leaves. Green color is important because it shows that there is plenty of Vitamin A. The hay should have at least 60 percent of its original color. It is especially important for your goat to obtain as much green roughage as possible if she happens to be pregnant. A lack of green roughage can result in weakened offspring due to a low supply of carotene.

If your feed dealer is willing to continue the discussion you began, you might ask him which stage of growth the hay was in when cut. You will soon discover that the quality of feed is determined by the protein content and that the more mature the hay was when cut, the greater the loss of nutrients. This will vary from the vegetative stage with about 19 percent protein to the mature stage when the protein content dropped to about 6.5 percent.

Lawn trimmings, garden trimmings, weeds, brush, leaves, etc., can all be utilized as roughage. You can even store these just as you would hay. When cut, spread them out and allow them to dry thoroughly in the sun, then fill your empty feed sacks and store. Just be sure that these trimmings do not constitute the total roughage. Your goat should have a good quality feed as well.

A word of warning about hay storage: do not store it where it can get wet. If this occurs, do not allow it to remain that way. Get it spread out and dried as quickly as possible, otherwise the mildew and mold will run rampant. Your goat will then refuse to eat it, and she will be smart to refuse, for it would give her a very upset stomach should she be foolish enough to do so. Don't overlook the possibility that this spoilage may have occurred prior to your obtaining the hay. If, when you open a bale, you suspect mold or mildew (the hay will stick together in little clumps of either whitish or brownish masses), don't hesitate to return it to your feed dealer. He should replace it with a good one without the least argument. Don't blame the feed dealer. After all, how is he to know what's in the middle of the bale unless his customers let him know?

GRAIN

Here you are on pretty safe ground. Pick a reliable manufacturer and purchase a bag of goat or calf "chow." This is a bagged mixture of all the various grains and ingredients needed to keep your goat up to par. If you are unsure about which brand to pick, ask your feed dealer's advice, look for yourself to see which bags have gathered dust on the shelf and which have not. Shelf life often tells you how good a product is.

Most states require that mixed feeds carry a tag. This will be found sewn into the top of the bag. It will list the ingredients and the chemical make-up of the feed. What you are looking for is the amount of protein and fat, which should be high; and fiber, which should be low. As a general rule, the fiber content should not run more than 12 percent and 8 percent is better. This concentrate should supply much of the needed protein. If you are feeding primarily a grass hay, the concentrate should be at a level of 16 to 18 percent protein. If you are feeding a

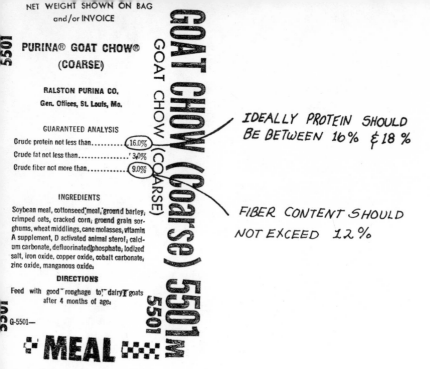

Sample feed tag.

legume hay, it may be as low as 12 to 14 percent. Again, cost is usually determined by these percentages.

If you continue to read these tags for several months you will undoubtedly discover that the formula will change from time to time. This is good. Manufacturers with flexible formulas do this so that they can continue to bring their customers the best quality for the same price. Just as with other foodstuffs, the ingredients that make up these mixtures will change in cost, according to season and availability. The manufacturers like to see their experts earn their money and will have them busily compounding new formulas to get the best for less whenever given a chance.

PELLETS

To me, pellets are an emergency measure to be used only in case the world suddenly runs out of fresh alfalfa and grain.

It is true that they are supposed to contain the correct ration of roughage and concentrate and all the vitamins and minerals that a goat should have. Of course, they are nice and neat and don't get loose hay all over the floor. However, goats are grazing animals by nature, and the pellets are somewhat similar to your taking a vitamin pill and expecting it to give you that nice full feeling that accompanies a good breakfast or comforting dinner. This is especially true when the most enjoyable pastime your goat has is to eat. If you take that away from her, she is bound to retaliate by breaking out in hives, or breaking out into your garden and devastating it. Pellets should certainly not be the total ration. Doing so will be certain to invite constipation problems.

VEGETABLES AND OTHER TIDBITS

Goats love fresh fruits and vegetables. They will adore the peelings from your dinner. However, don't tempt your goat with any fruits or vegetables that have been canned or frozen as these will upset her stomach. Just ask yourself, "Would she choose to eat it if it were growing in the garden?"

SUGGESTED MENUS

Now that you have all of the ingredients for a well-balanced ration, you will probably want to know how much of which to feed. This depends upon your goat: it is usually figured by body weight and whether or not she is pregnant or milking.

The smartest move you could make would be to check with your goat's former owner as to what he was feeding her. If these items are easy to obtain, not too costly, and she was happy with the combination, it would be foolish to change just for the sake of experimentation. Whatever you do, though, do it gradually, especially when changing from a less concentrated ration to a more concentrated one. When you ignore this rather important rule of feeding you will run into trouble with a constipated goat.

If you have purchased a goat and for some reasons are unable to discover what it was she ate before, or if she has informed you that what she has been getting at her last home was definitely not to her liking, and you have decided to reorganize her eating habits, the following are some fairly standard menu plans. Take into consideration that these are meant for the average goat. Decide for yourself whether or not your goat falls into this category, and tailor them to fit. Less is probably better than more, at least to start with.

Menu Number One
alfalfa or clover hay—3 pounds
grain mixture—1 to 3 pounds
Menu Number Two
alfalfa or mixed hay—2 to 3 pounds
chopped vegetables—1½ to 2 pounds
mixed grain—1 to 2 pounds

These amounts are meant to be the total day's ration, and should be broken down into two or three meals, depending upon your personal inclination. It is best to feed your goat the grain mixture while you are milking her, and give her the hay to relax with and mull over afterwards.

One rather important thing to remember is not to feed her any strong flavored vegetables (such as onions, garlic, or cabbage leaves) within four hours of milking. If you ignore this advice you will get milk that no one will appreciate.

All in all, your goat should consume approximately three pounds of roughage per day and one to three pounds of grain—one half a pound for each pound of milk produced. If you are feeding alfalfa, or other similar hay, a good way to judge about how much hay you are giving her is by noting the way a bale comes in sections. Each section weighs approximately five pounds. These sections are called flakes and can be separated into still smaller pieces. I plan on one flake per day. This gives me one pound to spill on the way to the pen, and one pound in case April wants a midnight snack.

SALT AND TRACE MINERALS

Goats need salt every day. A block of salt should be handy at all times. There is a nice selection of colors for you to make your choice from. There is the simple, ever-correct white, the sunny yellow sulfur which helps to discourage ticks and such, or the bright pink which will provide all those extra minerals she may be missing. Calcium and phosphorus are the two needed in the largest supply, followed closely by iron, copper, zinc, and cobalt. Just ask for a trace-mineral block if you don't wish to designate your choice by color. A word of warning: don't get too enthusiastic about stuffing your goat with all kinds of mineral supplements, even if you decide to use plain salt. She will probably receive plenty from the commercial mix alone.

If your goat is not partial to block salt, you might consider giving her the loose variety, adding a little to her grain each time. However, the salt block is a treat to most goats, so give her time, she will probably wear it down in short order.

WATER

Don't forget the all-important beverage, water. It should be fresh and handy at all times. Try to keep it in the shade in the summer. In the winter you might take the chill off by adding a little hot water. Your goat will drink more if it is made inviting this way. Remember, milk is mostly water.

FRESH GRASS

Mention was made in the previous chapter of utilizing your goat as a lawn mower. However, since that chapter did not deal with nutrition, no mention was made of how these elements will affect your goat.

Fresh grass is delicious to your goat, especially after she has been eating hay. Consequently, you might find your goat eating too much if she is suddenly confronted with an entire lawn.

If she consumes a great deal of fresh grass without prior warning to her system it may lead to founder (*see* Chapter XI).

Therefore, it would behoove you to let her out to graze gradually, increasing the time a bit more each day. Try half an hour the first day, working up to a full day over a period of a week or so. To keep her from gorging herself the first couple of days, feed her a little dry hay before turning her out.

Also, and it would seem obvious but I have seen it done all too often, don't stake your goat out in the blazing sun without some type of shelter or a shade tree to seek relief under. Remember to put out a pail of water, preferably in the shade. Look to see what flower patch you stake her near, and check the list of poisonous plants in Chapter XI to see that you are not tempting her with something lethal. If you have oak trees, watch out in the fall so that she doesn't find a batch of acorns. They are a tempting treat, but quite bad for her digestive system.

NUTRITIONAL DISEASES

While we are covering vitamins and minerals, it might be worthwhile to discuss the subject of nutritional ailments.

Nutritional deficiencies are brought about by too little feed, rations that are too low in one or more nutrients, or (sometimes) stress.

As for quantity of feed, if you have remembered to feed her twice a day, she should be fine in that category. As for quality, if you have fed her hay, a commercially mixed feed, and given her a salt block, she should be protected there as well. If you have not put her under any mental strain lately, she should be fine. So much for nutritional diseases.

PREGNANCY

During the last six weeks prior to kidding, the expectant mother will need slightly different nutritional care; not too much, just a little more than usual. When the kids arrive, there is a whole new field of nutrition awaiting you. This will be discussed in Chapters IX and X.

CHAPTER VI

GROOMING

Care and maintenance of your goat will be almost negligible. However, there are a few things that must be done to keep her healthy.

DAILY GROOMING

This is really a 50-50 proposition, being as much for your benefit as hers. It consists of taking a few quick swipes with a soft brush each time before you begin to milk. This brushing will accomplish three things:

(1) It will remove any loose hairs that might fall into the pail.

(2) It will keep her coat shiny and help stimulate good circulation, thus producing a healthy and happy goat.

(3) It will lull her into such a contented state that she will hardly know you are milking, thus helping to make the chore easier for both of you.

As for what type of brush to use, a soft scrub brush from the dime store is great. If you want something a little more fancy, you can pick up a "body" brush (the kind used for horses) at your feed dealer.

BATHING

This is an "iffy" kind of grooming job. If your goat is dirty, give her a bath. If she lives in a relatively clean pen, twice a year should do it. If she is a goat with many oil glands in her skin (as some humans have) she might need a bath every six weeks or so. Just don't overdo it, or you will probably end up with a goat with a dry skin and itchy problems.

She will need a bath more in the summer than she will in the winter. Unless she runs into a skunk, or similar catastrophe, I wouldn't advise washing her in cold weather anyway. Remember, goats instinctively know not to get wet. They catch cold too easily when immersed.

One way to tell if your goat needs a bath is if her milk suddenly starts having a goaty flavor.

For your bathing project find a nice sunny spot, free from drafts. You will need some inexpensive shampoo (special animal shampoo sometimes ends up being more expensive than human kind, and shampoo is shampoo); her brush (a good way to get it clean, too); a bucket of water, a hose, or more buckets of water, some cotton balls; and some soft towels.

First, put the cotton balls in her ears to keep out any water. Then wet your goat all over. Please check the water if you are using a hose. Make sure it is not icy cold nor coming directly out of a hose that has been lying in the sun—you will get nothing but trouble from your goat if you insist on starting off this way. Now, slather on some shampoo and, using your brush, work up a good lather. Keep dipping the brush into the pail of water to help accomplish this. Rinse her off and repeat the procedure. On the final rinse be sure to get out all of the shampoo. Now dry her thoroughly with the towels and brush her coat so that it all lies in one direction. Remove the cotton balls from her ears—you're going to wonder why she doesn't answer next time you call her if you don't. Tether her in a nice warm spot until she is completely dry. That's it!

HOOF TRIMMING

If your goat were out on the open mountainside, jumping from pinnacle to pinnacle, she would have little trouble with overgrown hooves. However, her pen and your lawn do not provide the rough surface needed to keep her hooves trimmed to the proper length. You must take care of this chore for her. About every six weeks will do nicely.

Like your shoes, your goat will probably wear her hooves out unevenly, or not at all. Therefore, when you notice that she is listing to one side, or has suddenly attained a greater stature, it is time to trim.

Again, no fancy tools are needed. It all depends on what you prefer. (After you have done this a few times, you will undoubtedly come up with your own favorite.) Some people use a pocket knife, some pruning shears, and some a farrier's knife. I found that a linoleum knife, which I had used to hack up the kitchen floor, worked out quite efficiently.

To begin with, this is not one of your goat's favorite pastimes. Therefore, you should ideally have two people involved—yourself and anyone else you can coerce into helping.

If however, you must do this job alone, secure your goat to some solid object and feed her, or otherwise entertain her. Don't take all afternoon doing this either, fifteen or twenty minutes should accomplish the job nicely.

Do her rear hooves first and get this over with. For some reason known only to goats, they hate having you touch their hind feet, but will allow you to play all day with the front ones.

Try to picture a blacksmith and you should have the correct position. For those who have never shod a horse, pay attention.

(A) Stand with your back to her head, pick up the foot and bend her leg so that the hock (see the anatomy chart in Chapter II) is between your knees. The bottom of the hoof should now be turned up so that you can work on it. Grasp the hoof in one hand and your cutting tool in the other.

(B) You will note that the outside of the hoof wall will have grown and turned under in a kind of scalloped effect. Trim this off first.

(C) Now, observe the soft inner area. This is called the "frog." Trim this as well, but work carefully while in this region or you may end up with a swift kick in the teeth.

(D) Continue to trim off the outside edges so that the hoof is level. Go easy and trim only a little at a time, or you may cut into the blood vessel located in the horny section of the

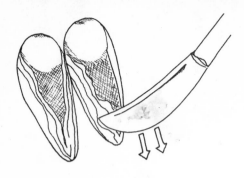

BEFORE

AFTER

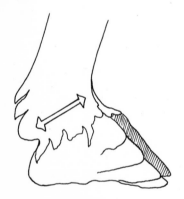
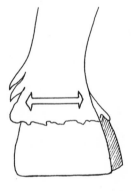

BEFORE

AFTER

Trimming the hooves.

hoof. If this should happen, you will know by one of two ways:
(1) a lot of blood will suddenly appear (2) your goat will bite
you on the back of the neck. Do not panic! Apply direct pressure on the cut hoof with a clean cloth for about thirty seconds,
disinfect the area, and get on with the job. Try not to repeat

this accident as it will tend to get your goat a bit upset, and she may not trust you for days.

If you take care to watch the color of the hoof and stop when it starts to get pinkish, you will avoid any unpleasantness in the first place.

(E) Now that the outside is nice and even, you may note that there is a decided lump at the back of the hoof, somewhat resembling the heel of a shoe. This also is part of the frog. Trim it too, remembering to stop if it starts to turn pink. That's as far as you can go today.

(F) Turn the hoof sideways and examine it in profile. It will probably resemble a triangle. This is wrong. Again, going very carefully, trim only the front part. The finished hoof should resemble a trapezoid.

(G) Repeat the process with the remaining three legs. When you move to the front, reverse the knee grasp—otherwise you will tend to dislocate your goat.

The best time to trim is after it rains as the feet are soft and pliable then. If necessary you can remedy dry, hard hoofs by soaking your goat's feet for a few minutes in some warm (not hot) water.

When you arrive at the point where you are raising kids, it is best to start them off right by trimming their feet at two or three months of age. They don't really need it much then, but it gets them used to the procedure. Be affectionate with them, make it a game, and you will have them begging you to give them a pedicure when they are mature.

CLIPPING

There are some goats who go through life without being clipped. They are happy, and their owners are happy. It depends somewhat on the goat and the circumstances. Some goats are hairier than others. Most goats in commercial dairies are kept clipped due to their occupation. If you are showing your goat you must clip her, as this is proper for the show ring.

I must admit that clipping is a chore that I would prefer not doing too often. I have thus hit upon a fairly good clipping program for my personal use. In the spring, after all the cold weather has definitely passed, and my goat is beginning to shed, I borrow my neighbor's poodle clippers and give her a complete once over.

You can use almost any type of animal clippers, from poodle to horse. Since clippers are very expensive, I would advise you to borrow a pair rather than invest in your own for your once a year clipping.

It doesn't seem to matter where on the goat you begin— one end is as good as another. This job will take you about one hour to complete: two, if you have to chase your goat much. Take frequent breaks, especially if your clippers are getting warm as this can be very uncomfortable for your goat. Use a half-nelson, or a cowboy throw, or whatever is easiest for you both.

In the summer months clip against the lie of the hair, and with it in the winter. This way you will have a shorter clip in the summer when it is hot and a longer one in the winter when it is cold. When you are going over a bumpy place, which will be just about everywhere, move the skin around and back and forth in order to get the clip as even as possible, and when you come to an area where the skin is loose, such as the elbows and stifle joints, stretch the skin as you clip so that you don't accidentally remove a chunk of skin. No one wants a goat with a lot of scars.

When you have finished clipping, brush all the loose hair out of her coat, and if your goat is not too annoyed with you by now, and you are not too pooped, you might consider giving her a short, soothing bath. If you do, be sure that she stays out of drafts until she is completely dry for she is now minus her winter coat, and if she is wet on top of that she could really catch a nasty cold.

That should do it for the summer season, When fall comes, and she begins to grow her winter coat, you can give her a partial clipping if you wish. This consists of clipping the flanks, udder, and belly. True, this is not the most glamorous of styles, but your goat shouldn't be planning a heavy social schedule for the winter anyway.

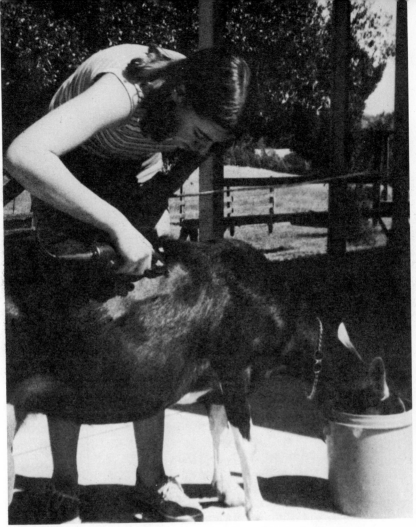

Clipping a goat is simple if you work quickly and keep her occupied.

DEHORNING

THIS IS NOT A DO-IT-YOURSELF PROJECT. Dehorning is normally done when the kid is from three to five days old, before the horns have started to develop.

If you have purchased a purebred or a recorded animal, it probably never entered your mind that she at one time sported horns. However, if you have somehow fallen heir to a mature goat that still has hers, read this carefully.

This is a job best handled by a veterinarian. Do not let your kindly old neighbor offer to remove those projections. This is somewhat similar to his offering to remove your wisdom teeth.

The complete dehorning of a mature goat is a major operation. It is very bloody and very painful and a severe shock to the animal; and if done when the goat is with kid, it is very possible that she may abort. Do not take it lightly.

One "popular" method of dehorning among laymen uses rubber bands. I shall describe this method, but I do *not* recommend it.

A notch is filed around the base of the horn and a strong rubber band is twisted and rolled into this grove. Adhesive tape is then wrapped around the horn to prevent the band from coming loose or being snagged on something. In most cases this must be replaced several times before the horn sloughs off in six to eight weeks.

The advocates of this method state that the pain period lasts *only* forty-five minutes to one hour. From having talked to a veterinarian, I understand that this is somewhat inaccurate and that the animal is in pain quite a bit longer, sometimes even for days. The veterinarian also said that there have been altogether too many cases of tetanus when this method is used.

Goats are normally dehorned for one main reason—to keep themselves from being hurt and from hurting other goats when living together in large groups. Goats like to play and butt each other, and if they do it with horns they can seriously hurt each other. If you plan to have only one or two goats and you are not going to show them, horns are not too critical a matter. And, if you want to take any precaution at all, you might have your veterinarian, or someone knowledgeable, saw them off about two inches from the skull. This should keep your goat out of trouble. Frankly, I think horns are kind of nice.

WATTLES

Wattles are little flaps of skin on the neck; somewhat like misplaced drop earrings. They are found on almost all breeds of goats. Like the human appendix, they are relics of evolution—remnants of gill slits, useful when your goat's ancestors were still swimming about under the reef. They usually are a nuisance only when the goats are young and the other kids use them as teething toys; they sometimes become infected as a result. Therefore, breeders usually remove them early in the kid's life. Show people say that their removal makes the neck look cleaner and trimmer, so there you are with the beauty angle again.

If you feel they must go, this is again a chore for the veterinarian. (Don't get the idea that I am pushing veterinarians, I just think they know the score and I trust them.) Some people remove the wattles by merely clipping them off with a sharp pair of scissors. I think you could liken this to having your ears pierced. I would rather go to someone professional for that, too.

If your veterinarian does the job, he will cut a circle at the base and remove a little wire-like projection which just cutting the wattle off with a pair of scissors cannot remove. If this projection is not removed, there is a definite possibility that a fistula will develop and then you've got a problem a lot worse than any pair of wattles could ever be.

Although I love my goats dearly, never could I in my wildest imaginings think of them as beautiful. What is one bump or lump, more or less, on an animal that already resembles a pile of leftovers from an anatomy class?

CHAPTER VII

MILKING

Once you get the hang of it, milking is really easy. However, unless you have had previous experience, plan to give yourself, and your new milk goat, about a week to get acquainted. In your first couple of attempts you will no doubt come rather rapidly to the conclusion that it takes more than just wishful thinking to make the thing work, and that Heidi must have been quite a girl. To get down to basics, here is the secret. It is as simple as learning how to parallel park a car; once you know you'll be set for life.

Pretend that you are going to shake hands; placing four fingers on one side of the teat and the thumb on the other. Start at the top, squeeze your thumb and forefinger together, shutting off the supply at the gap, so to speak. Then rotate downward with the rest of your fingers, forcing the milk out and into the pail. Loosen your thumb and forefinger and start all over.

The pressure should be firm but gentle. You will have to judge for yourself how much to use. Each animal is different. I found out that it took a little more pressure than I thought.

Be careful not to pull or stretch the udder as this could cause injury to the small blood vessels and you might just end up with blood in the milk, which is somewhat unappetizing and doesn't make your goat happy either.

For the first few days try doing only one side at a time until you get the feeling of it. Then you can go for broke and try for two. There is a definite rhythm to this. First you do one, then, as that one is refilling, you do the other. THINK. You have to release your thumb and finger on the one you have just done to allow it to fill again while you are squeezing the other. It takes a bit of practice, but suddenly you will get the hang of it and you'll feel like telling the world to come and watch. Don't! Your goat will immediately "dry up" and you will get nowhere.

This brings us to some of the finer points of milking. When I said that your goat would dry up, I meant just that. Goats are shy about milking. They get nervous when strangers, or the family dog, or your spouse, are standing around staring.

What you want is a lot of milk, right? To do this you must have a happy and contented goat. Remember that commercial about the contented cows? Well, it works with goats too.

There are a number of ways to make sure your goat will "let down" her milk. As I said, your goat must be happy. This means that you should reach a good rapport with her. Talk to her. Also, goats love music, so why not bring along your portable radio?

At the beginning of the milking session, while you are washing and drying her, massage her udder gently. This really works. Massaging the udder sends a nerve impulse very quickly to the brain. The goat will associate this with milking. The brain immediately sends an impulse to a nearby gland which in turn causes a stimulating substance to be injected into the bloodstream. The blood then carries this substance to the udder and the muscles in the udder are in turn stimulated. Your goat is now ready to do her part.

If you do not massage but go right to the milking, there will be a few little dribbles and drabs, but the milk won't really start to come down until a short time later. This is one of the reasons why I thought milking was such a chore when I first began. Then I got the message and began to massage. It only takes about fifteen extra seconds and is the key to effortless milking.

There. You've finished milking, or at least you think you have. Here is another little trick I learned by observing a professional dairy owner at work. When you think there is not another drop, again massage the udder, this time at the back, just for a second or two. You will get anywhere from three to twenty squirts more, which could be as much as a cup.

Most professionals end the milking with what is called "stripping." This is done by running your forefinger and thumb down the teat until you have gotten out the very last drop. They say that this is necessary for thorough milking, helps to keep up the goat's production level, and that the milk that comes with stripping is the richest milk in the udder—although I can't figure out how that one last drop can enhance two or three quarts of milk.

Now that you know the mechanics, let us take it from the top and go, step by step, through the entire procedure so that you will know the correct sequence.

It is morning. You know—it is an inescapable fact—that you must prepare to go out and milk that goat you bought in a fit of weakness.

First things first. What do you put the milk into? Well, it really doesn't matter very much. All you really need is any container that can be sent through the dishwasher; does not have seams where bacteria can multiply; is not made of a porous material such as plastic; and is large enough to contain at least three quarts of milk without spilling.

I finally settled on a ten-quart stainless steel Dutch oven that had a lid and also a handle for carrying. The lid allows me to cover the milk as soon as I have finished and helps to keep strange creatures from falling into it.

Having chosen a container for your milk, get a smaller one, any kind, and fill it with some warm, sudsy water or dairy disinfectant (this is a feed store item). Wash your hands and, tucking a roll of paper towels under your arm, head for your goat's quarters. Get her breakfast grain and set it before her. Wash and dry her udder, using the paper towels. (Cloth ones tend to collect germs.) Throw out the wash water and squirt the first couple of squirts from each teat into this container. This is to remove any bacteria that have crept into the openings since the last time you milked her. Also, you should examine this first milk for any lumpy or stringy substance. (If you find any, turn immediately to Chapter XI where you will learn all about udder disorders.)

Give this first milk to the dog or cat; they don't mind bacteria. Continue milking, this time into the "official" pail. Make sure not to get your hands in the milk. (I didn't really expect you to dip your pinkies in the milk, but sometimes they do get in the way of each other.) When you have finished, you put on the lid (hopefully you have one), say "thank you" to your goat, give her some hay, and head back to the kitchen where you will put the milk away. More about that in a moment, but first let us digress to take up the matter of where to milk your goat.

The place where you milk should be clean. By that I mean reasonably clean. Probably all the authorities would jump up in horror if they read this, but I milk April in her house. I kind of like the coziness of it, she likes it there, and we have a good time talking to each other. Goats are very clean animals and the bedding is nice clean straw and comfortable to sit on as I milk. April doesn't have to go out in bad weather, and I don't have to haul her to the porch, garage, or wherever.

Furthermore, sooner or later you will be told by just about everyone that has a smattering of knowledge about milk goats, that you must have a milking stand. This is a platform about six inches high with a yoke to put the goat's head into. It does three things for you:

(1) It does get the goat up higher so that you can sit on a stool instead of the floor.

(2) It holds her in place so that she can't get away.

(3) It takes up a lot of space in your garage, patio, or wherever.

I am like Thoreau: I hate to have one more thing I have to take care of. Therefore, item (3) carries more weight with me than either (1) or (2). Besides, I find that sitting cross-legged on the floor while milking helps keep me agile. And, since April is busy eating while I am milking, she is not interested in going anywhere until after she has finished.

Where and in what way you milk is something you must decide for yourself. Whatever method you choose, and whether your milking facilities are the pride of the neighborhood, or as makeshift as mine, it all comes out the same—and it all tastes great!

Now, back to the matter at hand. You have taken your milk into the kitchen and you want to know what to do with it.

Processing the milk is no great chore. Nor do you need any fancy equipment, although if you wish there is no end to the devices you can order through your dairy or farm equipment catalog, or from your feed dealer. I once browsed through the *Sears Farm and Ranch Catalog* and noticed stainless steel dairy pails, milk cans, a strainer, and filters. "Certainly I must need all of them," I thought, and proceeded to order the lot. What did they send me? Just the filters and a notice that the other items would be sent at a future date. Now, what does one do with a year's supply of filters and no strainer to put them in?

Filters are little round pieces of cotton material somewhat similar to the ones you use to make coffee with. (In fact coffee filters will do nicely in a pinch.) They are supposed to fit into the mouth of a large funnel-like contraption through which you strain the milk to get out anything foreign that has had the misfortune to have fallen in. This item is great if you are working with about twenty gallons of milk, but you aren't, so skip it. A plain household funnel will do the job as well. Just as you would prepare a filter for a filter coffee pot, shove the milk filter down into the funnel, put the funnel into the neck of whatever you are going to keep the milk in and pour away. Then throw away the filter. A year's supply costs about $3.50.

As for what to store your milk in, that is easy. If you want to be properly rustic you can buy a set of those clever little milk cans that are stainless steel replicas of the original five gallon farm kind. Besides being quaint, they really are quite practical. They will get your milk cold fast and keep it cold, and cold is what you want. However, they are expensive. The glass bottles that you buy juice in at the store work beautifully. So do large peanut butter jars, and they have screw tops that are easily cleaned.

In the matter of storing milk there are various schools of

This excellent milking stand was made from scrap lumber. Note the use of linoleum on the platform for easy cleaning and the recessed feed bowl.

thought. Most, however, agree on one point. To have sweet tasting milk you must get it cold quickly and keep it cold. This is important, too, in helping to keep the bacteria count down. Remember, you are dealing with raw milk. (More about pasteurization later.)

Authorities will tell you that the quickest way to cool the milk is by putting the container in an ice bath. This is simple enough. Fill your kitchen sink with cold water, throw in some ice cubes, and set the jars of strained milk into it. Leave them

Equipment for milking need not be expensive nor extensive. Shown here are paper towels to wash and clean the udder, a Dutch oven used as a milk pail, a choice of milk strainer or a kitchen funnel to be used in conjunction with the filters in the foreground, and a choice of an aluminum milk can or two glass jars for storage.

for about thirty minutes and then transfer them to the refrigerator.

Here is where the authorities split into two factions. One group says to store the milk without tightening the lids until after it is completely cooled. The other group says to tighten them and put them away. My thought on the first method is that raw milk, like butter, absorbs odors and your milk will soon taste like that leftover tuna salad on the second shelf.

From the beginning I did things the easy way. I brought the milk into the house, strained it into jars, screwed on the caps and tucked them away in the refrigerator. The milk tasted sweet and I was happy. However, I decided that I must try all methods to be fair and unbiased and I set up a controlled experiment.

I took the morning batch of five consecutive days and divided each batch into three separate groups. The first one I processed as usual and put directly away into the refrigerator.

The other two were put into an ice bath for thirty minutes. Both were then put into the refrigerator on the same shelf as group one. I tightened the lid of one and left the other to cool further. When I felt enough time had elapsed I tightened the lid on number three.

When all three had completely cooled, I had my kids (mine) and Herman (a finicky cat who refuses to drink milk that is less than perfect) act as judges. Not one subject could detect the slightest difference in any of the samples.

My conclusion was to return to my original method. You should probably try the various ways for yourself in order to find which suits you best. Perhaps your refrigerator has more exciting things in it than mine, or your ice cubes may not be as cold, or maybe it is a hot day. Factors differ in various situations, so whatever you feel most comfortable with is the best for you.

One thing to remember is that raw milk does absorb odors. Raw milk will also multiply in bacterial count faster at a warm temperature, and since the bacterial action will tend to change the flavor, you must take this into account as well.

As for the matter of caring for your milking equipment, most dairy books provide all kinds of sterilization methods. I am not putting them down; they are absolutely correct on sanitary matters. However, most dairy goat books are aiming at the professional or semi-professional who hopes to make money by selling his milk.

Now, I know you want to be healthy and not run a chance of introducing some strange germ into your family circle, but you do have to compromise somewhere. I don't think you have to sterilize everything, including yourself, and put on a fresh gown each time prior to milking. The following method is the one I adopted, and I haven't had so much as a cold in our family for over a year.

Decide what you are going to use for milking equipment and keep these items for this purpose only. Whether you are using a Dutch oven or a deluxe stainless steel milk pail with a hood and no seams, use it only for milking. Don't cook the pot roast in it on Thursday night, scrape it out on Friday morning and expect to get great tasting milk. It just won't happen. Keep your equipment clean and stored in as dust free a place as possible.

As soon as you have finished milking and have completed doing whatever you have decided to do with your milk, get right to your equipment care. Rinse the milk pail and funnel in cold water and put them in the dishwasher. Those dishwasher manufacturers know what they are doing—you will have a clean, sterilized pail the next time you go to see your goat about some milk.

If you don't have a dishwasher, you can do it the hard way. As soon as possible after milking, wash your equipment in hot soapy water and rinse with either hot water, or water to which a dairy disinfectant has been added. Let them drain dry. Do not dry them with a towel—germs multiply quickly in dish towels.

If you are really germ conscious (and you shouldn't be as goats are clean, healthy animals) you can even pasteurize your milk before putting it away.

PASTEURIZING MILK AT HOME

There are several ways of treating milk by heat in order to destroy the bacteria. Here are three of the easiest methods:

(1) Boil the milk over direct heat for three minutes, cool and refrigerate. This is easy. The only catch is that the milk tends to taste just like that—boiled.

(2) Heat the milk in a double boiler until the water in the bottom part has boiled for eight minutes. The milk will be a little better tasting than with method number one.

(3) The third method requies a dairy thermometer. (This is something like a candy thermometer; but don't try to use one of those as your milk will taste like the thermometer. You can obtain a dairy thermometer from your farm catalog, or from your feed dealer.)

Place your milk into a glass or stainless steel container. Put the thermometer in the milk: it will float. Heat the milk rapidly, stirring constantly with a stainless steel spoon until a temperature of 165°F. is reached. Hold at this temperature for twenty seconds, then place the pan immediately into a large pan of ice water, and with constant stirring, quickly reduce the temperature to 60°F. Pour the milk into jars and put into the refrigerator.

In addition to the processes above, electric pasteurizers for home use are available. These operate automatically and guarantee better temperature control than can be obtained by any of the above methods. Unfortunately, you do have to find room for them in your kitchen and you do have to clean them.

Better yet—drink it raw. Most people do.

OTHER THOUGHTS ON MILKING

Some things just will not fall into a logical order. The following are some of those little odds and ends that really should fit into a chapter such as this, but refuse to get into line. Hence the following:

One Sided Goats

When buying your goat, be sure to ask if she has been milked on one side only. Sometimes this may be the fact of the matter and if you don't realize it, you will not understand why she is not cooperating with you fully when you get her home. Decide now who is to be boss. If you are more comfortable on one side than the other, and I think it depends on whether you are a left-handed milker or a right-handed milker, let your goat know about it and let her adjust to you—not you to her. It will take about a week, but be persistent.

Our goat doesn't really care as long as she has plenty of food in front of her. My daughter is left handed and finds that she likes to milk from the right side, while I am just the opposite. Thus, we appear to have created an ambidexterous goat, for what it's worth.

Lying Down

Perhaps your goat continues to insist on lying down when you have decided that now is the time to milk her.

If this is her first time, perhaps the problem is simply that she does not understand the idea. Patience and gentleness will pay in this situation. Getting her acquainted with the operation instead of rushing the job will help to solve this problem.

Give her some grain, and while she is enjoying it, begin to gently massage her udder. Milk only enough to relieve her and try again later. Pretty soon you'll be going full tilt and so will she.

Don't overlook the possibility that perhaps she hurts somewhere. Maybe she has an injured teat or a cold in the head (check Chapter XI for possible symptoms). Goats usually like being milked.

Chapped Teats

Here is a really sore subject. (Sorry!) Winter is the time of year when the teats chap easily. These are best avoided by milking with dry hands, which you should anyway, and keeping her bedding clean and dry.

If this condition does occur, you can apply a bit of olive oil to your hands before milking. Use sparingly, though, or your milk will taste slightly of Caesar Salad. After milking apply a bit of vasoline with mentholatum in it.

Off-Flavored Milk

There are several reasons for a bad taste in the milk, all of which can be overcome with a little thought.

(1) Don't let your goat eat any strong flavored feeds within four hours prior to milking. Check the corners of her pen for any wild onions, etc.

(2) Bacterial action will occur rapidly in warm milk. Cool it quickly.

(3) Don't give milk a chance to absorb smells, and hence bad tastes. This can occur anywhere from the barnyard to the kitchen door. Cover your milk as soon as it is in the pail.

(4) Don't let any stray dirt fall into the milk, no matter how small. Again, keep the pail covered.

(5) If your goat is ill, or frightened, or has udder problems, the first place you will notice it is in the milk.

NOTE: Milk from a healthy goat will *foam* when it goes into the pail.

By eliminating these factors one at a time you should quickly be able to remedy any off-flavor that has occurred.

Hopefully, none of these problems will ever be yours.

CHAPTER VIII

COOKING WITH GOAT'S MILK

Since you probably now have much more milk than you normally would have on hand, you are probably wondering what to do with some of it—just as I did. Because I also have some chickens, custard was the first dish to make its debut. Even Quiche Lorraine found its way into the house occasionally. Through all of this I discovered some rather simple ideas that I thought I would pass along to you.

Since milk doesn't lend itself very well to main courses, there won't be any recipes in that category. You are, therefore, going to find mostly odds and ends like soups and desserts, a smattering of knowledge on butter and cheesemaking, and pointers on how to process your own canned milk at home.

BREAKFAST

There are two types of breakfast eaters—the light eater and the hearty, devil-may-care sort. For light eaters, here are two delicious drinks you can whip up in the blender. And, for the very timid, there is a terrific type of milk toast.

Copy Cat "Orange Julius"
Pour into the blender 1 cup of milk, ½ cup of orange juice, 2 tablespoons of honey, 3 tablespoons of instant milk powder, and 1 egg (optional). Add 2 or 3 ice cubes and whirl away until frothy.

Banana Boom

Pour into the blender 1 cup of milk, 1 small mashed banana, 2 tablespoons of honey, and ¼ teaspoon of rum flavoring. Blend well, pour into a mug, and top with nutmeg.

This is also good for husbands in bed with a cold and little boys just home from school. It isn't a bad idea for your lunch either.

Casper's Milk Toast

Take a couple of stale pieces of bread and make them even harder by toasting them in the oven. Break these up in a bowl. Meanwhile, slowly heat some milk. Pour this over the toast and add a dab of butter or sprinkle with cinnamon and sugar.

For those hearty souls, a good stick-to-the-ribs breakfast is just the ticket. Try one of these ideas, both guaranteed to make weak men strong and turn children into cherubs.

Double Whammy Cereal

On that cold, wintry day when you feel that it is your duty to make a pan of hot cereal for your family, use goat's milk instead of the water called for. Simple—but what a difference.

Double Whammy Eggs

Almost the same idea. Poach some eggs, but instead of using water, use milk.

Melt a little butter in the bottom of a frying pan. Break in as many eggs as you want, cover them with milk, put on a tight fitting lid, and simmer until they have reached the desired doneness. Add salt and pepper and put everything on some toast.

SUPER SOUPS, ETC.

Now we've taken care of breakfast, here are some ideas for lunch and supper. All of the following will fit equally well in either category:

Jolly Green Giant Chowder

Melt 3 tablespoons of bacon fat in a large pot, or fry 4 pieces of bacon that has been cut up. Dice 1 medium sized onion and ½ green pepper and add to the pot. Sauté until the onion is translucent, but not brown. Meanwhile, dice 1 large or 2 small potatoes and add to the peppers and onions. Add enough water to cover the potatoes and put on the lid. Cook until the potatoes are done; about 10 minutes. Add 2½ cups of milk and 1 can of drained whole-kernel corn. Add salt and pepper to taste. Heat through and serve.

The following recipe is really just a simplified version of fish chowder, but it is easy to fix and hearty enough for a Sunday supper, and the entire family should devour it with gusto.

Father Murphy's Chowder

Fry until partially done 6 pieces of finely diced bacon. Add ½ medium-sized onion and 2 large potatoes that have been finely diced. Add enough water to cover the potatoes, put on the lid, and cook until the potatoes are done. Add 2 cups of milk and 1 pound of cooked, flaked fish or 2 cans of minced clams. Add salt and pepper to taste. Heat through and serve with oyster crackers.

"Cheese It" Soup

Melt 2 tablespoons of butter in the bottom of a heavy pan. Sauté until translucent ½ medium-sized onion finely diced. Stir in 1 tablespoon of flour and slowly add 3 cups of milk, stirring all the while. Simmer this for about 5 minutes. Meanwhile combine 2 beaten egg yolks, ¼ cup of grated Cheddar cheese and ½ cup of cream. Gradually add this to the milk mixture and continue to cook for about 2 more minutes. Season to taste. Serve hot.

This next soup is a natural around Halloween, but it is also great year round. Don't go "yuk" until you've tried it!

Goblin Soup

Scald 3 cups of milk in a large pot. Stir in 1 large can of pumpkin (the kind you make pies with). It helps to use a whisk to stir in the pumpkin. When these two ingredients are well blended and smooth, add 1 tablespoon of butter, 2 tablespoons of brown sugar, and salt and pepper to taste. Heat through and serve.

If you want to be different, try this on company and add a blob of sour cream just before serving. Eight to five no one will guess what they are eating, and everyone will think it is terrific!

When this next recipe was devised it was close to pay day, but not close enough. About all that was in the house were some potatoes, a couple of onions, milk, and some cat food. (The cats got the cat food.)

Desperation Soup I

Melt some butter, or bacon fat, into the bottom of a heavy pot. Cut up and add one good sized onion. Sauté this while you are peeling and cutting up 2 large, or 3 small, potatoes into tiny pieces. Add the potatoes to the pot along with enough water to cover them. Add salt and pepper. Put the lid on and cook slowly for about ten minutes or until all the water is gone. Break up the potatoes somewhat. Slowly add 3 cups of milk and heat through. Serve hot.

If you have some bologna or salami on hand you can dice this up, fry it until crisp, and add it to the soup at the last minute. You can also sprinkle a little parsley on top when you dish it up.

If company's coming, you can put this soup through the blender, chill, and serve topped with unsweetened whipped cream and chopped chives. It is then called Vichyssoise—or pretty close to it.

If all you have on hand is onions and milk, try this.

Desperation Soup II

Melt about 3 tablespoons of butter (not bacon fat this time) in the bottom of a heavy pot. Slice 2 medium-sized onions into paper thin slices. Sauté these until they are a golden brown. Stir in 1 tablespoon of flour and ½ teaspoon of salt. Slowly add 4 cups of milk, stirring all the while. Heat well and serve.

The following recipe is not mine; nor can I remember from whence it came. All I know is that it really works, and I hope the original owner will forgive me if I pass it on. It is for cooks such as I who, when they find something foolproof and easy, treasure it.

Sneaky Souffle

Remove the crusts from 12 slices of white bread. Grate ½ pound of sharp cheese and layer the bread and cheese in a greased 3 quart casserole. Pour over this a mixture of 4 beaten eggs, 2⅔ cups of milk, ¼ cup of melted butter, 1 teaspoon dry mustard, ½ teaspoon of salt, and pepper to taste. Cover and store the casserole in the refrigerator for at least 12 hours—24 hours is better. Bake at 300°F. for one hour and serve immediately with a green salad. French bread, and that unpretentious little white wine you have been saving.

This next recipe is somewhat like the milk toast, and it's great for those quick suppers when it's just you and you don't feel like eating canned food.

Noodle Nosh

Warm up some milk, about 1 cup for each person you are going to serve. Get it to the simmering stage and add ½ cup of noodles (preferably the thin kind) for each 1 cup of milk. Cook until the noodles appear to be done. Ladle into soup bowls, add salt and some butter, and eat.

A SWEET FINISH

Some people simply must have their just desserts, and for those the following are included:

Custard is an obvious ploy when you are stuck with a lot of milk, but here is one that is slightly different.

Wiki Wiki Custard

Beat together 4 eggs, ⅓ cup of sugar, ½ teaspoon of salt, and 1 teaspoon of vanilla. Scald 4 cups of milk and add this to the egg mixture. Put 1 cup of shredded coconut in the bottom of a baking dish and pour the custard on top of that. Place in a pan of water and bake for 1 hour at 325°F. Cool. Top with whipped cream and slivered almonds.

Cowardly Yellow Sherbet

Stir together until dissolved 1⅓ cups of sugar and 7 tablespoons of lemon juice. Add to 3½ cups of milk. (Don't worry when the milk starts looking funny, it will straighten itself out when frozen.) Put this mixture into refrigerator trays and freeze. During the first hour, while it is still mushy, stir it around with a fork several times to break up the ice crystals. Continue to freeze for about 4 more hours. Serve.

Just-Plain Ice Cream

Beat 3 egg yolks and add ½ cup of sugar and ⅛ teaspoon of salt. Scald 1 cup of milk and slowly add to the egg mixture, stirring all the while. Continue to cook this until it reaches the boiling point. Cool and pour into refrigerator trays. Freeze for about 30 minutes. Meanwhile, beat until stiff 1 cup of heavy cream flavored with 1 teaspoon of vanilla. Remove the milk mixture from the refrigerator and turn into a bowl. Beat this until smooth and fold in the whipped cream. Return to the refrigerator trays and freeze for approximately 4 hours. Stir the mixture in the trays 2 or 3 times during the first hour to break up the ice crystals.

To make "just-plain" chocolate ice cream, omit the vanilla and add 8 squares of melted semi-sweet cooking chocolate.

To make "just-plain" polka-dotted ice cream, add 1 cup of small chocolate chips, or butterscotch chips, or both.

To make "just-plain" lemon drop ice cream, add 1 cup of crushed lemon drops.

Strawberry Yummy

Scald 2 cups of milk. Slowly add 3 beaten eggs and 1 cup of sugar. Cool. Add ½ envelope of plain gelatin which has been dissolved in ¼ cup of water. Whip 1 cup of heavy cream until stiff. Fold into the mixture along with 2 cups of sliced strawberries. Freeze in a refrigerator tray. Spoon into pretty glass dishes and top with one whole berry.

BUTTER

Making butter is really not too practical a project, but I think everyone ought to try it at least once. It is not hard. The main problem lies in that it is difficult to get enough cream to make any appreciable amount.

The cream in goat's milk rises very slowly and has to be collected a little at a time. It will take you about a week to collect enough. You can skim it off the top of the milk as it rises and freeze it a batch at a time until you have enough.

Defrost and place the cream in an electric mixer. Begin as if you were whipping cream. It will go through several stages rather quickly (much faster than cow's cream). At some point you will need to turn the mixer down so that you will not drench the kitchen in buttermilk.

Suddenly you will discover that there is a large mass either at the bottom of the bowl or stuck in the beaters. Scrape all of this together and run cold water into the bowl. With your hands make it into a lump and knead it until all the excess whey is removed. Return it to the bowl and let it sit under slowly running water for about 2 minutes. Salt it to taste and refrigerate.

Goat's butter will be almost white and slightly different, although delicious, in taste. If you wish, you can add some yellow food coloring in the final mixing.

CHEESES

Cheesemaking is something like winemaking. It must be done with care and know-how. In the case of wine, if you don't have the know-how, you end up with vinegar. In the case of cheese, you end up with a soggy mass of ill-smelling mold or something that could serve as a cornerstone for the Empire State Building.

To me cheesemaking is a true art. It is an art which I would like to master—given a year or two of spare time—but it is an art which I will have to forgo for the moment. You, on the other hand, may be chomping at the bit to get your hands on a cheese press and store up your cheeses to give to your aunt in New York City or that brother-in-law in Chicago who said you'd never stick it out in the suburbs. Fine! For you I will give directions for three common types of cheese. If you wish to pursue the art further, the library should be able to provide voluminous books on the subject.

Cottage Cheese

For this you will need a dairy thermometer. Place about one gallon (don't bother with anything smaller) of milk in a container of stainless steel or enamel. (Most recipes will tell you to use skim milk. Don't bother; you already know how much trouble it is to skim the cream off the milk. If you have made butter, you will have all that skim milk left over, and in that case, you're in business. If not, the cream will simply come out in the process of making the cheese.)

Place this pan of milk on the pilot light of your stove. If your stove is electric, you'll have to locate a similar area where you can keep it at a temperature of about 70°F. for about 24 hours. By that time it should be clabbered (thickened). When it is firm and smooth to the touch with a sort of watery substance (whey) floating on top, cut it into squares about the size of ice cubes. Now, put this pan into another pan that is half filled with water the same temperature as the curds. Heat this slowly until the curd reaches 100°F. Stir gently as it heats. Keep the mixture at this temperature for about 30 minutes, stirring every 5 minutes or so. Do not break the curds. They will settle to the bottom. The mixture is done when you can take a small piece of curd and squeeze it between your fingers with no milky substance issuing forth.

You should now take the curds, put them in a cheesecloth sack, and hang them in a cool place to drain. When the cheese has stopped leaking, but before it becomes too dry, transfer it to a glass or crockery dish and mix in about 1½ teaspoons of salt and a little cream, if you wish.

The taste of this homemade cottage cheese is hard to define. Unless you have tried it there is no way to describe it; but once you have, you will probably be hooked. It is nothing at all like the commercial type.

Neufchatel Cheese

This is a soft cheese that resembles cream cheese but is not so rich tasting—nor so fattening. It is great on toast or alone with fruit as a dessert or light lunch.

Again, start with one gallon of milk. Put it in a stainless steel bowl or enamel pan. Set this into a still larger pan of warm water. Bring the milk to 70° F. When both the water and the milk have reached the same temperature, add one rennet tablet which has been dissolved in ¼ cup of cold water. Stir thoroughly. Leave this alone for about 12 hours, or overnight. At this time there should be a firm curd on the bottom with a little whey on top. Pour off the whey and chill this separately. Put the curds into a colander and place a plate on top, weight it with something, and allow the curd to press together until dry. When the whey has chilled enough so that you can skim off the butter, do so and add it to the curd. Work this together until the cheese is smooth. Add about 1½ teaspoons of salt. Mix and store in the refrigerator in a closed container.

This cheese keeps only a few days, so use it up quickly.

Cracker Cheese

This is similar to the cheese you find in little glass jars at the store. It is a bit more of a bother to make, but if you think of the money you are saving, you will delve into this project with a surge of enthusiasm. This cheese is handy to have around and keeps quite a while in the refrigerator. It is good, as the name implies, on crackers or for lunchbox sandwiches.

Take one gallon of milk and allow it to clabber overnight. Heat the milk in the top of a double boiler to a temperature of 125°F., stirring occasionally. Put this into a cheesecloth sack and hang until dry. Add 3 rounded tablespoons of butter, a scant ¾ teaspoon of soda, 1½ teaspoons of salt and mix gently. Place the cheese in a glass or crockery bowl, press down so that no air bubbles are in the mixture, and set at the back of the stove for about 2½ hours.

Return the mixture to the double boiler. Add ¾ of a cup of sour cream and heat until the mixture has returned to the liquid state. Add bits of bacon or pimiento if you wish. Pour into jars, seal, and store in the refrigerator.

For a little conversation starter at your next party, put some of this out with crackers. After one and all have made their favorable comments, spread the news that this gourmet treat was a combined effort on the part of friend goat and yourself. If the comments are negative, remove the dish and keep the news to yourself.

HARD CHEESES

Here I must fail you. I have not included a recipe for hard cheese. There are many, but this book is meant to be a primer, not an advanced text. Nevertheless, if you get into this subject very far you will find cheese kits hard to resist. If you run across a foolproof method of whipping up a batch of Cheddar,

I would be grateful if you would share the secret with me. Just think of how much money you and I could save and what a feeling of accomplishment it would provide.

CANNED MILK

It is always a comfortable feeling to know that you have several jars of milk in case of emergency—such as when the goat steps in the milk pail just when you need it most.

Canning milk is really quite simple. You can put up three or four quarts while you are doing the morning dishes, or whatever it is you do in the morning.

Strain and cool the fresh milk to 50°F., or less. Pour it into clean glass canning jars up to within an inch of the tops. Put the lids and bands on and screw them down tightly. Place the jars in a large kettle that is partly filled with water. Cover and bring the water to a boil. Let it boil for one hour, being careful that it doesn't boil down too far. Cool and store in a dark place.

FREEZING MILK

If you have a freezer with a lot of empty space, this is a great place to store that extra milk. Goat's milk lends itself to quick freezing better than cow's milk. Use any home freezer container. Pour in the milk (allowing for expansion) and freeze away.

MISCELLANEOUS DATA

For what it's worth, milk, in dairy goat language, is always quoted in pounds instead of quarts and gallons. The reason for this is that milk varies in weight, depending on its composition. The richer in butterfat it is, the less a quart of milk will weigh. If you keep at this sort of thing, you may run across a recipe that calls for pounds and such. However, for all practical purposes a pint of milk can be considered to equal one pound, or eight pounds to the gallon. Remember, "a pint's a pound the world around."

CHAPTER IX

BREEDING, PREGNANCY, AND BIRTH

It is one of those indisputable facts of nature that in order to make more milk, you must make more goats. Therefore, at some point you must consider breeding your goat.

I am going to presuppose that the goat you have brought home to share your lawn is already supplying you with milk and that you are interested in keeping the supply going. The current lactation period will last anywhere from eight months to a year or more, but the average is ten months, which gives her two months to dry up and get ready for the next cycle. As you will notice, this fits nicely into the span of a year.

Your mother-to-be will come into season from about September on into the winter, although some anxious ones will come in as early as August. Since it takes about five months to produce the offspring, this brings their arrival nicely into the late winter or early spring months—just when the world is reawakening from winter, and the young kids will be weaned in time to catch the new grass on the rise.

Perhaps you like to do things from scratch and have purchased a young goat who has never been a mother. The proper age at which to breed a youngster such as this will depend mainly upon her individual development, but as both you and she are novices, it would probably be best to wait until she is at least eighteen months of age.

If she comes from a line of particularly heavy milkers there is a possibility that she may be what is known as a "precocious milker." This means that somewhere around the first year of age her udder may begin to develop, and even without the help of motherhood she may come into milk. Great! Begin to milk her, but still plan on breeding her at about eighteen months.

THE BUCK

If your goat is a purebred, it is a fair assumption that you got her from either a breeder somewhere in your vicinity, or a person who will know where to go to find a breeder of your goat's type.

In the first instance, you need look no further: call the breeder and set the date. On the other hand, if you would like to do a bit of shopping, you might check the list of associations in the back of this book and check with them for breeders to be found near your home. Also, don't forget your friendly Farm Bureau or local 4-H or FFA leaders. They can help put you in touch with someone who has an excellent buck of good lines and a low stud fee.

Stay with a buck of the same breed as your goat rather than another. In other words, don't try crossbreeding. Remember that recorded grades, which is what this mating will produce, go for considerably less than purebred kids. If you have a reasonably nice looking goat, and you breed her to a buck of some distinction, you can expect to get anywhere from $125 and up for your doe kids.

You should plan to pay an average of $15 to $25 for the stud fee. Here, again, I am presupposing that you do not intend to breed for show, but are just looking for a nice substantial mate of average build and intelligence.

If you have either a recorded grade or a scrub goat, you will still be better off breeding her to a purebred buck. As noted earlier, the closer you get to your breed in looks, the more you can expect to ask for your kids. If you have a recorded grade that is say a cross between a Nubian and an Alpine, which is fairly common, you would be wise to pick a buck of the breed that your goat comes closest to in appearance. Usually the ears tell the story.

Champion Laurelwood Acres Activator—an excellent Toggenburg buck. (Photograph by Bill Serpa; courtesy Laurelwood Acres.)

With good recorded grade kids, you can expect to ask anywhere from $75 and up, depending on their quality.

LINEBREEDING

Linebreeding is a polite term for incest among animals. It is one of the best methods known to produce superior offspring. Your breeder will give your goat the once-over and know exactly what characteristics to emphasize and which to elimi-nate. If any defects or weakness occur in the linebred strains, linebreeding can amplify these as well as the good points, but your breeder will know not to push these defects. Remember that the French Alpine breed of today was started with only four does. Somewhere, somehow, there had to be some line-breeding, and look at what a great breed resulted.

ARTIFICIAL INSEMINATION

Just so you won't go away wondering what you missed, let me give you a word about A.I. (pro talk for what we are dis-cussing).

As an individual owner, it seems more than likely that you can find a suitable buck somewhere within driving distance. So, unless you are looking for some unusual strain, I would suggest that you stay away from this method.

The rules for registering youngsters produced artificially with either the American Goat Society or the American Dairy Goat Association are quite stringent, and the costs involved are certainly no bargain.

THE MATING

It is late summer and your goat should be getting excited about the coming event. Some of the signs of coming into season are bleating, nervousness, falling off in milk production, and a frantic wagging of the tail. Once she has started to come into season, she will continue to do so about every twenty-one days. Thus, if you breed her and she does not show evidence of coming back in, you can be fairly sure that the job has been done. You should now figure 150 days from the day that the mating occurred. Check the following guide if you have trouble with figures. It should give you a close approximation of when to expect the arrival of the kids.

Gestation Table

Aug. 4	Dec. 31	Nov. 2	Mar. 31
Aug. 9	Jan. 5	Nov. 7	Apr. 5
Aug. 14	Jan. 10	Nov. 12	Apr. 10
Aug. 19	Jan. 15	Nov. 17	Apr. 15
Aug. 24	Jan. 20	Nov. 22	Apr. 20
Aug. 29	Jan. 25	Nov. 27	Apr. 25
Sept. 3	Jan. 30	Dec. 2	Apr. 30
Sept. 8	Feb. 4	Dec. 7	May 5
Sept. 13	Feb. 9	Dec. 12	May 10
Sept. 18	Feb. 14	Dec. 17	May 15
Sept. 23	Feb. 19	Dec. 22	May 20
Sept. 28	Feb. 24	Dec. 27	May 25
Oct. 3	Mar. 1	Dec. 31	May 29
Oct. 8	Mar. 6	Jan. 1	May 30
Oct. 13	Mar. 11	Jan. 6	June 4
Oct. 18	Mar. 16	Jan. 11	June 9
Oct. 23	Mar. 21	Jan. 16	June 14
Oct. 28	Mar. 26	Jan. 21	June 19

If the days drag by from May to September and your goat does not show any of the signs of coming into season, you might discuss this matter with your veterinarian. He can give her an injection which should bring her around. Also, I understand that the use of Vitamin E capsules sometimes gives things a boost, but I'm not going to quote anything without conferring with the FDA.

While we're on the subject of medication, about one week prior to breeding your goat be sure to give her an inoculation against Enterotoxemia and Tetanus. There is a combination injection now on the market of Clostridium Perfringens Type D and Tetanus that will fill her with antibodies so that she can pass these on to her offspring and keep them happy and healthy until they can receive their own supply at about three months of age.

The actual breeding is a quick affair and should only take a few minutes of the buck's time. However, there is one small catch. It is extremely likely that your goat can go into season and out of it again without your even realizing its passage. Sometimes her season can be a matter of only a few hours. If you have been out shopping or over at a neighbor's having coffee, you could miss the whole affair. Therefore, it might be wise to make an arrangement with the breeder to board her with him for a few days when you feel that the time is approaching. He will place her in a pen close to the buck, and then when she comes into season, he will let nature take its course.

At this point, you can take her home and make plans for the arrival of the offspring.

CARE OF THE PREGNANT GOAT

Some chapters back we discussed feeding of the pregnant goat. Of all the areas of caring for the expectant mother, this would seem to be the most important, especially when you remember that not only is she producing milk for your table but is also busy producing two or more offspring as well. It goes without saying that your goat should be given the proper rations to make her happy and healthy.

About two-thirds of the birth weight of the developing young is attained during the last six weeks of her pregnancy. As a result, her protein requirement is about one-third higher at that time and you should increase her concentrate allowance accordingly, along with a good quality roughage. She should be getting about one pound of hay daily for each thirty pounds of body weight. Minerals, calcium, and phosphorus are necessary for bone and muscle development, so you may want to consider adding some extra calcium at this time, but do check with someone knowledgeable first.

A few days prior to kidding about 50 percent of the grain should be replaced by wheat bran, which will act as a mild laxative. It is important that your goat not be constipated at this time.

You should also wait for at least one week after kidding before putting her back on her full grain ration, and should then do so gradually. In general, feed her a good bulky mixture of concentrates. A good one is equal parts of oats and bran.

DRYING UP

Your goat will need about six to eight weeks of rest prior to kidding. She should be allowed to "dry up" (stop giving milk) so that she can store up nutrition for herself and her offspring.

In drying up, there are two different methods for you to consider. It all depends on the temperament of your particular goat and whether or not she happens to be a heavy milker.

If she is not, you can try the following method: Milk her out completely, reduce her grain to about one pound per day for approximately one week, and feed her a great deal of dry roughage. When she has completely dried up, increase her grain gradually to the level that it was before you started this regime.

On the other hand, if she is a heavy milker, you may run into problems if you try the method given above. I am a coward and will sit staring at a full udder waiting for it to burst into a million pieces. Therefore, try the second method and the coward's way out.

Omit one milking per day for three or four days, then decrease to every other day, and so on, until things have taken care of themselves.

If, at any point in your drying-up schedule, you notice too much bulging or if she is in obvious discomfort, milk off enough to relieve her, but no more.

Once in a while you will run into a goat that will never dry up completely. Usually these are the ones that were precocious milkers. If you are so fortunate, or unfortunate, as to have one of these, you should accept this abundance gracefully, but try to taper off at least three days prior to the kidding to allow the colostrum to build up in the milk.

EXERCISE

There are times in the life of a pregnant goat when she just doesn't feel like making the effort. It is necessary, though, for her to get her blood circulating in order to feel really up to par. This is especially true on those cold days when she would prefer to lie abed.

If you are lazy you can make her get her own exercise by spotting her food at intervals about her abode. That way she will at least have to walk about in order to eat. However, if you feel up to it, there is nothing like a brisk walk about the neighborhood with your goat at heel.

GIVING BIRTH

As the time draws near, you are nervous, you imagine the worst, you anticipate every moment, you look deep into those great brown eyes, and . . . she shrugs her shoulders. Goats are smart, but not that smart. If it makes you more comfortable, you might act on the following suggestions to make yourself feel more useful.

Clip the areas around her udder, hindquarters, and tail.

Make sure that her bedding is clean and dry each day.

Make sure that she has plenty of fresh water, and that the bucket is placed out of the way so that she will not drop a kid in it accidentally.

If the weather is rather bad, you might put a bar across the doorway to keep her from having her kids in the elements.

Your goat is an easy birthing animal, should need little or no assistance, and will probably wait until you are well out of the way before she gives birth.

She may have from one to four kids, with five and six being rather unusual. (If that occurs be sure to call your local newspaper and claim celebrity status.)

Signs that the time is near include a rising tail bone with a falling away of the skin around the base of it and a looseness around the exterior genital area, rapid cud chewing, restlessness, plaintive bleatings, rapidly filling udder, and a mucous discharge which is first transparent, then opaque. These symptoms pretty much come in order, or there may be no warning at all. You may just have gone out to feed her breakfast, proceed to water the lawn or take out the trash, and return to pat her on the head—only to find a couple of new arrivals.

Most likely the best thing you can do for your goat is to leave her alone. If anything really unusual should happen, RUN, don't walk, to the phone where you should make a fast call to your veterinarian.

Your goat will probably clean and care for the kids as they are born. Give her a chance to do this before you interfere with your dirty, germ-ridden hands. If she totally ignores them, you can wipe the mucous from their mouths and nostrils and dry them with a soft, clean towel. Then, in either case, paint the kid's navels with iodine. Fill a coffee cup with 7 percent iodine solution (the kind you buy at the feed store) and dip the navel right into this solution. More is better than less.

The normal presentation is with the two front feet coming first with the nose lying between them, and in most events this will happen before you can say, "they're here."

If all goes well she should expel the afterbirth about half an hour or so after the last of the kids is born. However, it can take much longer. Do not become impatient and try to pull it out. Nature must do that, not you, since to do otherwise may cause serious internal injuries.

If you plan to have the veterinarian come in for a post-partum check-up, try and save the afterbirth for his inspection. He can tell from looking at it if all has been expelled and everything is all right.

Some goat mothers are inclined to eat the afterbirth. This sounds repulsive, but it is a leftover instinct from her days in the wilds when the smell of an afterbirth would alert predators to the birth of the kids. If this happens, the most serious effect seems to be a lack of interest in food for a day or two, and this can be taken care of with a little warm water and bicarbonate of soda or an Alka Seltzer or two.

There will be a normal bloody discharge for some days afterward. This discharge should be watery and not bright red. If it is the latter, check with the veterinarian as it is a sign that something is wrong.

Following her labors, give the mother some warm water to

which you might add a cup or two of molasses; and a bran mash made with equal parts of oats and bran which have been steeped in hot water and allowed to cool to a nice tummy-warming mixture. Give her a fresh supply of hay and turn your attention to the kids.

THE KIDS

Most breeders and professionals on the subject will follow the procedure of removing the kids from the mother immediately after birth. The mother is so busy having them that she is not supposed to know that they are there. I don't think goats are all that dumb. Also, I feel that it is best not to go against nature, and that the mother is really the best individual to care for the kids during those first important days of life, at least until you are a more experienced hand in these matters. I also find that if they are allowed to stay with their mother they are better adjusted mentally as well as physically. The mother will get over that first surge of motherhood and become bored with them after a few weeks anyway, and you can *gradually* wean them away from her at that time. Then, by the time they are ready to go to new homes, things will be settled with all parties concerned.

The whole point of this adventure was for you to get the milk, right? Here is the secret: you do not let the kids nurse. Actually most of them can't find the teats anyway. As soon as you know that the birthing was successful and that all have been accounted for, you milk the mother just enough to give the kids their first meal.

If you start right from the beginning to feed the kids from a bottle, they will never catch on that the stuff comes from the mother, but will consider *you* their meal ticket. After three or four days you can switch the kids over to a formula of canned milk and water, or if the mother goat has more than enough, use that.

You may begin to use the milk for yourself about the third day after kidding. It is a good idea to note the milk as it enters the pail: when it foams freely, it is good for drinking.

It is of prime importance that the kids receive the first milk which contains the colostrum. This is a thick, yellow milk packed with vitamins, antibodies, and a laxative substance which will help the kids expel the black, tarry substance that is the first feces. You should be sure to watch for these feces as much as you did for the afterbirth and, if nothing happens, call the veterinarian for further instructions. Give the first milk a chance to work. However, about twelve hours is about the limit. Just keep track of what is in the straw; you don't have to stand around waiting.

If for any reason the colostrum is not available, give the kids some warm milk with a teaspoon of Milk of Magnesia and hope for the best.

To feed the kids, use sterilized soft drink bottles and nipples that you get from the feed store. Make sure that the milk is at body temperature, just as you would for any baby.

CHAPTER X

RAISING THE KIDS

One of the amazing things about newborn kids is that they are so well coordinated at the moment of birth. They are up and jumping about in a matter of minutes, sizing up their new surroundings, and checking to see what's for dinner.

However, once in a while you will run across a kid, who, through no fault of his own, was folded into a corner inside the mother and got his legs into an awkward position. This kid will take a day or two to straighten out, but give him time before you rush him off to the veterinarian to be put into casts and traction units.

One of my own prize-winning senior kids spent the first two days romping around the pen on her knees. By the following week she was outdistancing her sister in the broadjump, and has since won a sizeable number of blue ribbons. So, sometimes it pays to take a "wait and see" attitude.

WHAT'S FOR DINNER?

Those first few days are rough. You must count on feeding the kids about five times a day for the first three days. That means that you will have to set up a schedule of about one feeding every five hours. Don't plan on any long social engagements unless you can hire an understanding babysitter.

Using soft drink bottles and the nipples that you bought at the feed store, plan to give each kid about four ounces at each feeding, increasing the amount gradually to eight ounces by the end of three days.

Now, don't go measuring things "just so." Use "about" measurements. Some kids eat more, and some are kind of picky and have small stomachs. The average soft drink bottle contains sixteen ounces, so use that to gauge amounts.

After three days you should be able to cut the feedings to three a day, and can even change over to a formula if you wish. This can be either canned evaporated milk mixed half and half with water, or a milk replacer that you can obtain from your feed dealer.

From three days to two weeks you can work up the amount to about twelve ounces at each feeding. About this time the kids will start investigating their mother's food. This is all to the good and they will be easier to wean if they get to liking the adult fare.

At two weeks of age you can cut the feedings to twice a day. At this point, the kids should be consuming a full sixteen ounce bottle, and perhaps a smidgen more.

Until they are two or three weeks of age the only liquid they will consume is milk, but at about that age they will suddenly trip over the water dish one morning and decide to give it a try.

Occasionally you will find a non-conforming youngster who has figured out where the milk is really coming from. This one will wait until your back is turned and then nurse for himself. Be a little smarter: smear the mother's teats with a blob of mentholatum, or other ill-tasting concoction. This should get him in line in a hurry.

If in the first few weeks any of the kids begin to scour (have diarrhea) you should not waste time in treating it. This can be serious. Try cutting the milk in half by diluting it with water; or change over to skimmed milk and boil it first, then cool it, and give it to the kid. It is important that they not lose any more body moisture than you can help. You might also add a little Kaopectate to the formula. It couldn't hurt!

And, if the opposite is the problem, put a dose of Milk of Magnesia in the milk, using the same dosage as indicated for a baby of the same weight.

I realize that the following advice is probably superfluous,

One-month-old Alpine doe kid already showing the characteristics of a good future milker.

An Alpine Senior Kid Doe. Next year she will probably be a good mother as well as a first-year milking doe. (Photograph by Victor Baldwin.)

but there may be a bachelor or two out there who has never done this sort of thing before. When feeding with a bottle, be sure that you have the milk at body temperature. You know, test it on your wrist just like they do in those family movies and to feed, tip the bottle up at a 45° angle so that the kid gets milk, not air. I am not about to tell you how to burp a baby goat.

WEANING

You are going to have to decide for yourself at about what age you plan to sell, or adopt out, your brood. Breeders will take the kids away at birth and begin to auction them off as young as three days. However, if you don't know what you are doing, the mortality rate is high. My personal plan is to send them packing at around six weeks to two months. At this age they are fairly well weaned, have a good lease on life, and there is no great strain on the part of either your goat or the kids.

To get started off on the right track, begin at about two weeks of age to give the kids or your goat an outing on the other side of the house, or anywhere where she can't hear them bleat and they can't hear her bah-ha. Give them both some hay to distract them, and don't give in too easily. Start with about ten minutes and gradually lengthen the time to several hours, or even the entire day.

However, do leave the kids with Mom long enough so that she can teach them the rudiments of eating hay and grain and such.

Weaning them away from the bottle is somewhat like the drying-up process. When you have them down to two bottles a day and they are around six weeks of age and growing like weeds, drop the morning bottle for about one week, and then one cruel night, skip the whole thing.

TATTOOING

Tattooing is basically for pedigreed goats. So that the people in the American Goat Society or the American Dairy Goat Association will know who's who, they will assign to your individual purebred (or recorded grade) kids letters and numbers that are to be permanently scripted on the insides of their ears. These numbers will also be listed on the registration papers so that everyone else will know who's who as well.

It goes something like this: If Mischief Acres is the name of your farm, then MA will be assigned to you to use in all of your kids' right ears. Then, to even things up, you will use the letter that has been assigned to the current year ("K" is being used for 1976) and add to that the number of the kid as born that year, i.e., K1, K2, K3, and so on, and put that in the left ear.

Tattooing is done with an instrument that resembles a stapler. It is equipped with a head that has little needle points that form the letters and numbers. Indelible ink is rubbed into the area on the inside of the ear and the tattoo is applied. It does hurt somewhat, but it only takes a second, and they forget it the next. Don't put it off. It is something that should be done concurrently with registration so that you can tell which kid is which without guessing.

Most dairy farms, 4-H project leaders, and FFA leaders can direct you to someone who will be glad to either loan you this equipment or even do it for you.

DEHORNING

The earlier this is done the better. There are two methods to choose from.

The chemical method can be done by you and is best done when the kids are about four or five days old. This is a salve that you can obtain either from your feed dealer or veterinarian. Apply it to the area where the horns are starting to erupt, but take care not to get it on anything else, including yourself. It is quite caustic and will burn very quickly. The application of this ointment will be rather painful and the kid may try to rub it off. Don't let this happen or you will have to start all over again. Be sure that the kid doesn't get it in his eyes as it can blind him.

The other method uses a disbudding iron. This is similar to a branding iron with a little round head. It is heated until it is red-hot and is applied to the bud of the horn for about one minute.

This is not a job for anyone with a queasy stomach or too much empathy. You probably don't have a disbudding iron hanging around the house anyway. Your veterinarian will do this job for you at a very nominal cost.

INOCULATION

No one seems to be immune (sorry) to inoculation programs, including the ruminant member of your family. After all, why keep a sick goat roaming about your backyard?

Starting with kidhood, your goats should receive inoculations against Enterotoxemia and Tetanus at about ages twelve and sixteen weeks and as yearly boosters thereafter.

Learn how to give these yourself. Have your overworked veterinarian give you a lesson, go home and practice on an orange, and then play doctor yourself. The serum can be picked up at most feed dealers for pennies and your goat will never know what bit her if you let her have a cup of grain while giving her the inoculation. Letting your veterinarian give the inoculations will cost more, but if you are squeamish I guess it is worth it.

WATTLES

Remember those wattles we spoke about in a previous chapter? Well, removing them is strictly a matter of preference, but should you decide against them, check with your veterinarian at the time you are having the horns taken care of, and leave it to him.

BUCK KIDS

Let us hope that you have been blessed with two or three doe (girl) kids. However, if somehow fate has led you astray and has presented you with one or more buck (male) kids, you must do something with them. No one wants a buck, with the exception of a doe, and then only for a moment. Their destiny thus is to be either someone's pet or someone's dinner.

Either way, they must be castrated. Castration is something that should be done only by a professional, be it veterinarian or breeder. The cost is minimal and can be absorbed in the sale of the weather (castrated male kid). You can expect to get from $15 to $25 depending on the buyer and the kid. The castration will eliminate any developing goaty odor, and if you plan to keep the kid and raise him as a meat animal, it will allow him to fatten.

I must in all openness discuss this possibility with you. Goat meat tastes very much like lamb, so they tell me. Goat meat is sold by reputable meat dealers under the trade name "chevon," and by some not-so-reputable dealers as lamb. There is no Federal restriction against marketing it in this way. Thus, it is entirely possible that you have already tasted a "leg of goat" and not known it.

The decision is one I will leave to you. I shall most likely end up with a pasture populated with goats named Harvey, Harold, Justin, Lawrence, Henry, etc.

CHAPTER XI

ILLNESS

Before delving into symptoms which I hope your goat will never exhibit, let us discuss the prevention of illness. We have discussed good nutrition and good housing, and if you have been following this advice you should have a healthy, stalwart goat ready to stand firm against the attack of any germ that comes her way. There are, however, some parts of the prevention program which are necessary in addition to commonsense and cleanliness.

WORMS

Worming, as part of the prevention plan, has been left to now to discuss. All grazing animals are subject to internal parasites, whose life cycle is usually to be found partially in the ground and partially in the animal. Here is where the single, or two-goat, family is somewhat in luck, for unless you have had a large flock of sheep on your front lawn, your goat will probably not be too heavily infested.

Some breeders suggest following a regular worming schedule of twice a year, spring and fall. I personally don't think you should overdo worming. Too much is probably even more harmful than too little since the medication is toxic and could kill your goat instead of a spare worm or two. In other words, treat only when you suspect your goat has become infected to the point where she should be treated.

The signs of worm infestation are pale membranes in the

mouth and around the eyes, rough coats, soft droppings, and thinness. Don't confuse thinness with the hat-rack appearance that a healthy milk goat should have. Adult goats usually develop a certain amount of resistance to internal parasite infection.

The effects of worms are mainly anemia, damage to the stomach and intestinal linings, and interference with digestion and the absorption of food.

Numerous books have been written on the subject of worms, with long Latin names and much medical jargon attached. What you really need to know is whether or not your goat is presently infested.

There are several species of worms and you probably should be aware of the different types, if nothing else.

The blood sucking worms all live in the stomach and intestines. They enter the animal via infested grass. Among this group can be found the brown stomach worm, thread-necked worm, hairworm, and whipworm. There is also a hookworm which should be included in this group, but it differs in its method of entry. Along with being ingested by mouth, it can also enter through the skin. All these worms are somewhat alike in appearance. They are hairlike and vary from ⅓ to 1½ inches in length.

The lungworm has a similar cycle in that they are ingested and pass in the feces; but after entering the intestines they proceed to travel to the lungs where they can cause a great deal of damage if infestation is heavy. Animals with lungworms usually have a deep cough.

Goats can be infested by several species of tapeworms. These are long, flat worms about ¾ of an inch in width and somewhat resembling a cooked noodle, both in color and shape. They can attain a length of several yards, and are usually found in the small intestine of the animal. Most authorities feel that the common tapeworm is fairly harmless. Their presence does not produce any particular symptoms, and the goat does not appear to be affected. Tapeworms require a different type of medication than the one for common stomach worms. If you feel strongly about this, I would suggest talking it over with your vet or feed dealer. But, if my goat has one, it's her business and I am not going to pursue it further.

"A cigarette a day keeps the worms away" is a favorite saying of some oldtime breeders. I am not quite sure of this as an everyday practice, but I understand that nicotine does do something bad to the worms. However, don't rush out and buy your goat a pack of cigarettes, it might do something bad to her as well.

The medication used to eradicate most worms, other than tapeworms, can be purchased from your feed dealer. You have merely to tell him that it is for a goat, her approximate weight, age, and whether or not she is pregnant. This medication comes in several forms and several flavors. Prices seem to go along with the palatability of the product. If you have a choosy goat, be prepared to buy the expensive stuff. If not, you can pick up whatever is cheapest.

Your goat probably has a pretty sharp eye and an even better equipped nose. To just hand her the medication and expect her to eat it is a bit foolish. There are, therefore, several methods of sneaking it into her. The ever-popular molasses ploy is my favorite. The heavy odor and sweet taste disguise the medication quite successfully. Simply make a little candy ball and present it to her. Another method is to leave the medication in the freezer overnight and then casually mix it with her morning grain ration just before serving. If you discover a better method, let me know.

EXTERNAL PARASITES

Now that you have the horrible feeling that your goat is filled with squirmy things, we can discuss the things that can infest her exterior.

Goats are extremely clean animals and external parasites should not be a problem, but you really should know about some of the most common ones.

MITES

Mites are the cause of the disease known as mange. They are tiny parasites, barely visible to the naked eye. There are several species and each type of animal has its own kind. Thus, if you have a dog with mange, it is most unlikely that your goat will contract the disease. Let us hope that you do not have a dog with mange anyway.

There are two forms of mange. One form is caused by a type of burrowing mite; the other by a type of mite that bites. If the mange is severe enough, milk production may be affected. The disease usually occurs in the winter when the animal spends much of her time inside.

The symptoms are scabs on the skin and loss of hair. The mites can be controlled by an application of a topical ointment available at the feed store.

Lice

Lice are also a winter problem. They are of two types: those that attach themselves to the skin and suck the blood, and those which bite and live on the scales. They spend their entire lives on the goat. You can treat lice with sodium fluoride. This should be mixed with equal parts of sulfur and lime and dusted into the coat. When a goat scratches herself, it can be a symptom of lice. Lice do not live in oil, and animals that are healthy usually have enough oil in their skins to make it impossible for lice to exist. Note: There is a clear-cut warning implied here. If you are looking for a goat to purchase and you spot lice, pass her by.

Ringworm

Ringworm is a contagious disease of the skin caused by a microscopic mold or fungus. Small round red areas develop, along with some loss of hair. These patches occur mainly in the head and neck area. Treat the affected areas with iodine every three days until the infection clears up.

Screwworms

One of the most undesirable external parasites is the screwworm. Fortunately for those in the North, the screwworm fly is found only in the Southern and Southwestern states. It deposits its eggs on the surfaces of an open wound where the eggs hatch and dig into the skin. There they feed and grow for about a week, drop off, and bury themselves in the soil to develop into the adult fly. If the damage done to the tissue is extensive enough it can cause death. If you practice common fly prevention and watch for infestation in any open sores, you should have no problem.

DETECTING ILLNESS

Once you are familiar with your goat, you will be able to detect if there is something wrong with her fairly readily. However, just because she is suddenly taking naps at three o'clock instead of one o'clock, you should not take this as a sign to call the veterinarian. On the other hand, odd behavior that doesn't seem quite normal can be a sign that you should start watching her just a little more closely to see if anything further develops. My one and only experience of illness with April happened just this way.

April, by her nature, is a bah-hah-er. By that I mean that she bah-hahs a great deal of the time. (This is a bad habit, I know, but I dote on her and I hope that she will continue to bah-hah for many years to come.) One morning, though, I noticed that something was amiss: it was much too quiet in the direction of April's pen. I went out there and looked at her, but other than her silence, she appeared to be her normal self. I decided that she had merely chosen to have a quiet day.

The next morning she was still silent, and I began to worry. I felt rather foolish calling the veterinarian, but he didn't think so, and had me bring her up posthaste.

She had a really bad case of founder and it was only my

gut feeling and knowledge of her usual habits that roused my suspicion. Henceforth:

WHEN IN DOUBT—CALL THE VETERINARIAN

And, when you do call the veterinarian, probably one of the first bits of information he will ask for, just as your family doctor would, is your goat's temperature. Which brings us to what you should have in your goat's first-aid kit.

To begin with, find a strong box of some type to put the things into. It really is important that you have a separate box for all the things you will need in case of an emergency, since when you need that thermometer, you don't want to have to search through the woodworking tools in order to find it. What follows is a basic list of items to put into this box. You will no doubt add to it as you go along, with an eye ointment here and a sulfa powder there; and in the end what you have will probably resemble a well prepared dispensary, but for now let us start with the basics:

a thermometer—animal type with string attached
scissors
tweezers—handy for everything
disposable plastic syringe—for giving medicine
roll of gauze bandage
adhesive tape—any size will do
sterile cotton in a roll
Vaseline
wound disinfectant—any type, preferably non-stinging
eye wash

As for first-aid, goats are pretty much the same as any warm-blooded animal. If they are injured, keep them quiet. If they are bleeding, try using a pressure bandage to stop the flow. If they are not breathing, try artificial respiration.

If you suspect illness, or poison, or a similar catastrophe, check her temperature as well as the other vital signs.

The normal rectal temperature for a goat averages 103.8°F., with a range of 101.7°F. to 105.3°F. To take her temperature, insert the thermometer into her rectum and leave it there for a minimum of three minutes. Use a lubricant such as Vaseline. It is important to have a piece of string tied around one end should you get too ambitious about its placement and need to retrieve it in a hurry.

In general, infectious diseases are usually associated with a rise in body temperature, but the temperature can also fluctuate with the weather, exercise, excitement, age, feed, and so on. Therefore, do not use this as your sole gauge. The body temperature is usually lower in colder weather, at night, and in older animals.

The normal pulse rate is 70 to 80 beats per minute and can be taken by feeling the inside of the thigh where the femoral artery comes closest to the surface. As with the temperature, the pulse rate is also affected by exercise, nervousness, age, and so on. The more exercise or excitement, the higher the rate; and conversely, the older or more sedentary the animal, the slower the pulse.

The breathing rate ranges from 12 to 20 inhalations per minute. It can be found by placing your hand on your goat's flank and feeling the rise and fall, or by watching her breath on the air in cold weather. As with humans, exercise, excitement, and hot weather will speed up the breathing rate. This is only normal. However, pain will also cause rapid breathing. So, if none of the other factors are present, you should definitely consider this a possibility.

Before we discuss the various ailments your goat may have, it might be wise to mention the ways to administer medication.

If the medication is a solid, such as a tablet, you merely have to place it on the back of her tongue and hold her mouth closed until she is forced to swallow. If it is a liquid, such as Milk of Magnesia, one rather successful method is to utilize the plastic part of a syringe used for giving injections (back to the feed store again). Draw up the amount of liquid to the correct mark on the barrel of the syringe, place the end in one corner of her mouth and slowly push in the plunger, a little at a time to allow her to swallow, until all the liquid has been consumed.

A goat is a hardy creature and will usually go through life

with no trouble. However, once in a while they will succumb to the attack of a bacteria or two.

As in any family medical reference book, there are several diseases in the following listings that will not occur too often; they are there because you really ought to be aware of their existence. Also, you just might have a hypochondriac goat and may have to reassure her that what she does have is merely a cold in the head and not an attack of coccidiosis. These listings (in alphabetical order) are here mainly for reference when you feel that your goat is looking a bit ill.

Agalactia (Suppression of Milk)

This usually occurs at kidding time. It is caused by such factors as indigestion, loss of appetite, mastitis, insufficient or unsuitable feed, plant poisoning, thirst, or undue excitement.

The goat should be put in soothing and comfortable surroundings and given plenty of fresh drinking water and good hay. A warm bran mash may help to stimulate the milk flow. Massaging the udder with ointment may also help to bring about the milk secretion. In any case, efforts should be made to milk her at least twice a day, if not more, until normal function has returned. WHEN IN DOUBT—CALL THE VETERINARIAN.

Anemia (Nutritional)

An iron deficiency will usually be the reason for anemia, but it can also be caused by a lack of copper, cobalt, or certain other vitamins and minerals. There will be a loss of appetite, progressive emaciation, and sometimes death. It is seen most frequently in nursing kids. The treatment and prevention are the same. Provide dietary supplements for the lacking nutrients. A trace mineral salt should always be available. If you are feeding a good commercial mix, good hay, and providing a trace of mineral salt block, there should rarely be reason to suspect this problem.

Bee Stings

Bees are attracted to darker colors. This is not something to worry about too much, however, as goats, like other animals, develop an immunity and will show little reaction to the stings. The exception to this is a badly stung udder: there you are in trouble! CALL THE VETERINARIAN.

Bloat

This is similar to colic in horses. It is an accumulation of excessive amounts of gas in the stomach. It is dangerous and often fatal due to pressure against the heart. If you suspect this, and it is pretty easily spotted, don't wait, get her to the veterinarian quickly.

The symptoms are a sudden distinct swelling in the triangle formed by the left hip bone, the end of the rib cage, and the top of the loin. The goat may show distress by lying down and getting up, pawing at her stomach, and making sounds of pain.

Bloat may result from overeating of tender, young grass, especially when it is still wet with dew. It may occasionally be caused by eating vegetables and apples that have not been cut up.

It is best to prevent bloat by making sure your goat eats some dry roughage prior to being turned out on new grass.

If you are having trouble contacting the veterinarian, an emergency measure is to give her an Alka Seltzer or some club soda via a syringe. Be careful not to force any more down her than she will take, or you may just do more harm than good.

When you do get her to the veterinarian, he will probably introduce an antiferment via a stomach tube and possibly even puncture the stomach wall with a hypodermic needle to allow

the gas to escape. Don't wait. THIS IS A TIME FOR QUICK ACTION!
EMERGENCY!

Bronchitis

Remember that cold in the head. This is it. The symptoms are
a running nose and watery eyes, shivering, dull coat, quick
breathing, and a dry cough.

A vaporizer, if you have one, would be a great help. I am
not above putting some Vicks on her chest and nose. Find a
nice, old warm sweater and make her comfortable. Tempt her
with a bran mash and keep concentrates down to a minimum.
If her breathing doesn't sound too good, or she starts to look
bad, CALL THE VETERINARIAN.

Brucellosis (Malta Fever)

This is a serious disease, but since the lesions which cause
it are not often evident, there is not much you can do about it.
The main symptom is abortion. Not all abortions are caused by
this, of course, but this could be a reason. In any case, if your
goat should abort, you should always call the veterinarian.

If it helps to place this disease in your mind, it is known
as undulant fever in man. Count the number of people you
know with undulant fever, and you'll have a rough idea as to
the possibility ratio of your goat having malta fever.

Bursitis

Goats spend a lot of time on their knees. Sometimes an older
goat will get a swelling around the kneecap of either one or
both knees. This might be bursitis, or inflammation of the knee
joint. If you take her to the veterinarian, he will perform a
simple operation that should take care of it. He will insert a
needle into the swelling to allow the fluid to drain and then
replace this with an equal amount of a solution of one percent
silver nitrate. The knee will be painted with iodine and your
goat will be ready to go home, as good as new.

Coccidiosis

A microscopic protozoan which inhabits the intestinal tract is responsible for this disease. The symptoms include extreme thinness, lack of appetite, and diarrhea which is often blood streaked.

If your goat has any of these symptoms, you should certainly contact the veterinarian. He will probably examine a sample of her droppings to see if any organisms are lurking there. More than likely, he will then treat her with a sulfa drug or antibiotic. Careful nursing is subsequently required on your part to bring her back to her old self.

Constipation

This can occur with goats of any age. Hard droppings and straining are symptoms. If a goat refuses her grain at any time, it should be considered a possible signal that all is not well. Withhold all further grain and reduce the amount of roughage. You may offer her some laxative feed such as a bran mash. Milk of Magnesia is a good medication and is easily administered. You use the same amount of medication per pound of body weight as you would for yourself.

When your goat is again up to par, increase her grain ration gradually; you might consider trying a different brand just to see if that has any effect. Perhaps you should also check the hay you are feeding for large amounts of coarse, dry, indigestible matter, as this can also be a cause.

Diarrhea (Scours)

In adults, this can be a symptom of disease, but it can also be caused by a bit of bad feed or just an upset stomach. If your goat appears to be happy otherwise, try giving her a dose of Pepto Bismol with the syringe. Again, use the same dosage for her as you would for yourself, poundage wise. Cut out concentrates and give her free choice of roughages. If she doesn't rally around soon, check with the veterinarian. He will probably want to see her and will offer stronger advice.

In kids this is termed scours, and it can be quite serious. It usually develops during the first few days of life if it is going to develop at all. The kids that tend to scour are usually the weaker ones. They will show a lack of wanting to nurse and will be depressed as well. Scours is usually the result of an infection caused by cold, wet, unsanitary conditions, and prevention is therefore rather easy. Treatment at home is similar to that for humans. Boil the milk, cool it, and try to get the kid to nurse. A dose of Kaopectate added to the milk may help. If the condition does not stop soon, you had best call the veterinarian. Do not wait too long: kids with scours can die of dehydration in a matter of hours.

Encephalitis (Circling Disease)

This is an infectious disease occurring most often in winter and spring and is nearly always fatal. Animals of all ages are susceptible. It is caused by a bacteria called Listeria monocytogenes which invades the central nervous system. The disease is quick and death is rapid. It is similar to sleeping sickness. The symptoms are uncoordination, circling (thus the name), and paralysis. Because this disease is an infectious one, it occurs mainly in flocks—which one or two goats cannot be classified as, so don't lose sleep over this one.

Enterotoxemia

This is an acute disease and death comes rapidly. It is caused by a type of anaerobic bacterium called Clostridium perfringens, Type D. It is found in the soil and in the digestive tract of nearly every warm-blooded animal. Overeating of lush feed seems to bring on the disease.

There is a rather quaint saying among veterinarians that, "A sure symptom of Enterotoxemia is rapid death." Prevent this from ever happening to your goats by making sure that they are inoculated against this disease both in kidhood and in yearly boosters.

False Pregnancy

If you have bred your doe and she appears to be pregnant, but time passes and she fails to produce a kid, you should consult your veterinarian. It is entirely possible that instead of motherhood, she has a sick uterus. Your veterinarian will cure the problem and you can try again next year.

Fluorine Poisoning

The ingesting of excess quantities of fluorine from water or feed is the cause of this ailment. Fluorine is a cumulative poison and can be found in concentrated quantities in certain areas. You should be aware of it if it is in your soil. There is certainly no secret as to what parts of the country are awash with it.

The symptoms are stiffness of the joints, loss of appetite, abnormal teeth, reduction in milk flow, diarrhea, and salt hunger.

You should remove any sources, such as local water, and avoid the use of feeds or mineral supplements containing excess fluorine. (The upper limit is 0.01 percent of the total dry ration.) Any damage already done may be permanent.

Foot Rot

This is the athlete's foot fungus of the goat world. It will not occur if you keep your goat's quarters clean and dry, as it is a symptom of sloppy housekeeping. The first signs will be lameness followed by a swelling of the foot which becomes hot to the touch. Check with your veterinarian for a good remedy, most likely an antibiotic of some sort.

Founder

Most people tend to think of founder as something that happens only to horses, but it is common in all hoofed animals. It is caused mainly by overeating of concentrates, or over-drinking when hot. If you take care to store the grain where she can't get at it and take care not to think that "the darling needs just a little bit more," because of the way she is looking at you with those large bovine eyes, you will not be agent to the cause. She is not going to overdrink if you make sure that there is fresh water available to her at all times.

There is also a type of founder that is caused by the retention of a part of the afterbirth following parturition. Hence, when your goat kids you should save the afterbirth to show the veterinarian so that he can determine if all is where it should be.

In founder there is a high fever, reluctance to move, and extreme pain in the legs. Quick action should be taken. Wrap her legs in towels that have been dipped in cold water and keep her quiet. This is a serious and painful disease; take quick action. CALL THE VETERINARIAN.

Goat Pox

This is similar to cow pox, if that tells you anything, but is caused by a different virus. At any rate, it is certainly an ugly looking disease. When it first appears there are small, red, swollen areas which cover the surface of the udder. These areas will be about the size of a pea, but will soon grow larger. They will quickly come to a head and eventually break open, spreading pus everywhere, and leaving small raw sores behind.

Treatment consists of painting each sore with a tincture of iodine and then applying carbolated vasoline to the healthy areas of the udder. This will prevent fresh sores from forming and eventually the pox will disappear—possibly within a week, possibly longer.

Drinking the milk is not dangerous unless the infection has entered the teat itself. In that case discard the milk until all is well. If you feel insecure, check with your veterinarian.

Goiter

Goiter is caused by a failure of the body to obtain sufficient iodine. The best prevention is to keep a trace mineral salt block within reach at all times. The symptom is a large swelling in the neck, and once this appears, no treatment is effective. Just keep your goat supplied with salt, and you shouldn't have to worry about goiter.

Ketosis

This is a serious disease and occurs just prior to or shortly after kidding. An unbalanced diet or any sudden changes are considered to be the cause. Prevention is the best answer. Watch carefully to keep the hay and grain ratios consistent and try to feed at regular hours. Do not become obsessive over this, however.

The first symptoms are muscular spasms and loss of appetite. As the disease continues, coma will develop along with rapid breathing, and finally, death.

The treatment is intravenous glucose and intestinal stimulants. But, as stated, the best treatment is prevention.

Mastitis

This is an inflammation of the mammary gland. Usually only one side is affected. It will probably be hot and painful to the touch. If the disease is really established, there may be a discharge that is either thick and yellowish or thin and reddish. The animal usually has a rather high temperature as well.

An injury to the teat can allow for the bacterial invasion which causes mastitis. If you suspect any injuries, it is best to care for them before they develop into this disease.

Your veterinarian should be consulted if you suspect mastitis. He will probably suggest using hot packs made from towels soaked in Epsom salts. Massaging the udder with camphorated oil after it has been milked out is also helpful. The vet may also suggest injecting the udder with an antibiotic. This usually comes in a small, plastic, individually packaged syringe arrangement which is inserted directly into the teat through the natural opening. This may sound complicated, but it really isn't, and surprisingly the goat won't mind very much either.

Sometimes, as a result of this disease, lumps will form in the udder, some as big as golf balls. They can be reduced in size by massaging gently after milking. However, the udder will never again produce the quantity of milk that it did prior to the infection.

Milk Fever

Milk fever is a definite misnomer as there is no actual fever. This disease occurs in the heaviest producers and appears soon after kidding. It is caused by a lack of calcium. Unless it is treated promptly, it is most likely to be fatal.

The symptoms are loss of appetite, dullness, twitching of the muscles, excitement, and restlessness, followed by coma and death. CALL THE VETERINARIAN QUICKLY.

Your vet will probably inject her with calcium salts intravenously. If she is treated in time, she will begin to respond rapidly, usually within a few minutes, and be her old self again by the next day. Again, prevention is the best cure. Don't stint on her feed, especially in the green roughage department.

Osteomalacia

This ailment is rather rare. The symptoms are a depraved appetite (eating bones, wood, and other similar objects) or conversely a lack of appetite, stiffness of joints, and decreased milk production. The cause is a lack of vitamin D, inadequate intake of calcium and phosphorus, and/or the incorrect ratio of calcium and phosphorus. Give your goat sunbaths, a trace mineral block, and don't worry.

Photosensitization

This occurs primarily in sheep and cattle, especially the all-white varieties. It occasionally happens to a goat. Only the white areas are affected. They become hypersensitive to sunlight after eating certain plants such as leaf horsebrush, spineless horsebrush, smartweed, St. Johnswort, and blossoming buckwheat. I don't know your garden that well, but it's a pretty sure bet that you don't have these mixed in along with your crabgrass.

The symptoms are swelling of the ears, eyelids, and lips, along with intense itching; somewhat similar to a severe sunburn. Sometimes these swellings form scabs. A soothing preparation should be applied, and the animal kept in the shade.

Pneumonia

It is entirely possible that your goat, or your neighbor's goat, may be a carrier of this disease and never become ill with it. The disease usually develops after chilling, poor feeding, and overexposure in bad weather. The symptoms are fever, labored breathing, and lack of appetite. If you suspect more than just a slight cold, you should call your veterinarian. He will probably give her sulfonamides or antibiotics.

Pregnancy Disease

This is more a lack of good nutrition rather than a real disease. It usually occurs in late pregnancy and in a doe with multiple fetuses. Grinding of the teeth, weakness, frequent urination, dullness, and trembling are a few of the signs. If you feed her a good balanced ration there should be no problem. Should you suspect that your goat is carrying several kids, you might add some molasses to her water. It can't hurt and sometimes it really makes the difference.

Salt Deficiency

By now you probably have salt blocks lined up ten deep in order to prevent a lack of trace minerals. Therefore, you should never have to concern yourself with your goat's having a salt deficiency. Just be sure to have salt available for grazing animals at all times, especially in the summer.

Some of the symptoms of a salt deficiency are loss of appetite, loss of weight, retarded growth, rough coat, and low milk production.

Sometimes the goat will get stubborn and will refuse to spend enough time licking her salt block. In that case you might try mixing some loose salt into ther grain ration. This type of salt can be obtained at the feed store and looks like the salt you use to make ice cream.

Tetanus

This is mainly a wound infection that attacks the nervous system of horses and man. However, it does occur in goats, especially if you happen to have horses in the same area. It is caused by an organism which is anaerobic and causes trouble when it enters a deep puncture wound. DO NOT TAKE CHANCES. Have your goat given a vaccination with tetanus toxoid when you purchase her, if her former owner has not already done so. If you neglect to, and she does sustain a wound that is suspect, short term protection can be obtained with use of tetanus antitoxin, which should be given as soon as possible. If the wound is bad, she will probably need one anyway, even if she has had the other. I think this is where that ounce of prevention comes in—just make that ounce one of tetanus toxoid.

Tuberculosis

This is one of the touchy points that might have bothered you when you were considering the purchase of a milk goat. "What were all those things you always heard about the constant testing of cows for tuberculosis, and raw milk being a scary thing?" Here is what the experts have to say on the subject: "Tuberculosis among goats is generally regarded as a rare disease."

This disease is usually contracted by eating feed or drinking water contaminated by the discharges of infected animals. (How many infected cows have stopped by your house lately?) A symptom of the disease is coughing. If your goat does start coughing, it is more than likely that it is a cold in the head rather than tuberculosis.

Udder Injuries

If your goat should injure one of her teats, or udder, so badly that milk is flowing through the wound, you should take her to your veterinarian and have him suture the wound as soon as possible. Under most circumstances, a sterile milking tube can be inserted so that the animal can continue to produce milk while the injury is healing.

If your goat shows no outward signs of injury yet shows clots of blood in the milk, or pink tinged milk, it is possible that there may be a ruptured blood vessel within the udder. This happens most frequently in animals that have large pendulous udders.

Most of these problems will clear up by themselves if you will milk out only enough from the injured half to relieve the pressure each time. If the injury is severe, the veterinarian may use a drug to hasten clotting, injecting it intravenously, or into the udder.

If you have nursing kids, they can sometimes be a bit rough. Try taking them away for a few days and milk her by hand until the condition clears up.

Warts

Warts are a rather common condition in cattle, sheep, and goats. They are caused by a virus and will spread if not stopped. An application of carbolated vaseline should help to clear them up, and castor oil is also a good home remedy. The length of treatment depends on the individual goat. Some cases disappear almost immediately, others seem to persist for several weeks.

MISCELLANEOUS PROBLEMS

When you run across symptoms that don't seem to fit any special disease, don't overlook tooth troubles. Your goat can get them too. This is especially true if she slobbers, is off her feed, or has bad breath. Abscessed teeth will cause the goat to run a temperature. Check with your veterinarian, since he is your goat's dentist as well.

Also, don't overlook the possibility that your goat may be suffering from an allergy. Just because she can chomp her way through a stand of poison ivy it doesn't mean that she won't break out in hives when offered some green apples or peanuts.

POISON PLANTS

Which brings us to plants that are definitely poisonous to your goat. Most animals are pretty selective about what they consume and will not eat a poisonous plant unless they are actually starving. Nevertheless, your goat may be so happy chewing her way through your vegetable patch that she will never notice that she has just taken in a fatal portion of oleander until it is too late. It is a wise owner who knows his own weed patch, and I am listing some of the plants that can poison your goat.

It is really amazing how many common plants you may have in your garden that can be harmful and even fatal to your goat. Some of these are rhododendron, azalea, mountain laurel, braken fern, buttercup, cowslips, water hemlock, yew, foxglove, delphinium, lobelia, lily-of-the-valley, and oleander—which is about the most deadly. Even the smoke from its burning leaves can poison a pasture. Oleander is the shrub they use frequently in the center dividers of freeways, and that is the only place it belongs as an ornamental plant. It can kill your toddler just as effectively, so watch out.

It would seem sensible not to allow your goat near any of these plants, but there will be a day when your goat will refuse to pay attention. Goats do not eat tin cans, but they are sometimes strange in their appetites. If April escapes her pen, the first place she will head is for the roses. You would think that the thorns would tickle going down, but she can clean a path through my rose garden in less time than a herd of aphids. I have one friend whose goat loves to get at his cherry tree. She will consume her fair share and then, reclining on the front porch, burp up the pits one at a time and spit them at the dog who is usually sleeping nearby.

One thing that no one would think of is potato poisoning. In the spring when new potatoes are just coming in, the green coloring that is sometimes still on the peelings contains a substance that affects the nervous and/or digestive systems. It would be wise not to give your goat the peelings from these new potatoes.

Toadstools are, of course, a foolish thing to give your goat, but perhaps you might miss a couple when you gather the grass clippings. Just don't overlook this possibility, for your goat will probably gobble them up right along with the clippings, and there you are—no goat.

It is also a good idea to check the labels of the various fertilizers and insecticides you are using on any foliage or grass she may munch on. Naturally, snail and slug bait pellets are deadly.

If, after all the precautions, you find yourself with a goat full of poison, my first suggestion would be to take the label, sample, or plant with you and run—not walk—to your veterinarian.

The usual signs of plant poisoning are vomiting, frothing at the mouth, staggering, and convulsions, along with cries of pain.

Goats have a lot of things going for them, and one is a complicated digestive system. Drenching and intravenous treatments are the best methods of combating poisoning and are best done by the veterinarian. If, however, you cannot reach your vet, and you have tried calling several others, you might try a purge of Epsom salts.

Put 4 or 5 tablespoons of Epsom salts in some warm water, and using your plastic syringe, squirt it into her mouth, letting her swallow it each time before you give her more. Be careful that you are not strangling or drowning your goat in trying to help her. If she cannot swallow, it is probably because her throat is paralyzed, and you should not try to force anything down her at this point, for you will only succeed in killing her faster. Also, if her stomach is distended, use your own common sense about putting anything else in there.

If what has poisoned your goat seems to make her drowsy, try using a drench of strong tea that has been quickly cooled. Tea is three times more effective than coffee. Again, be especially careful about putting down too much, too fast. She will probably have a hard time swallowing. A teaspoon at a time is best. Small doses of liquor will sometimes help as well. If she appears to be going into a coma, try rubbing her all over briskly with a towel that has been dipped in cold water.

Summer will bring with it another form of plant menace. When the natural vegetation of spring begins to dry in the heat of summer, vicious by-products such as foxtails, burrs, and other prickly forms of seed pods will sometimes work their way into the udder and other unprotected areas of your goat's skin. The udder is especially vulnerable as it is usually moist with summer heat.

You will first notice a small lump which will soon begin to abcess. You can usually dispose of this problem yourself. Clean the area with alcohol, and using a pair of tweezers, extract the matter as you would an infected splinter. Clean the wound and dust with powdered disinfectant to help keep the area dry, speed healing, and prevent more of the same.

Forewarned is forearmed, and armed with the list of symptoms above I am sure your mind will be at rest in the knowledge that whatever your goat has, it is probably not fatal.

A foxtail or two in summer and a case of the sniffles in winter is about par for the course, so put your mind at rest and go on to better things upon which to dwell.

> "May there always be sweet milk in your pail
> and a taste of honey forever on your lips."

APPENDIX

BREED REGISTRY ASSOCIATIONS

American Angora Goat
  Breeders Association
Mrs. Thomas L. Taylor,
  Secretary
Rocksprings, Texas 78880

American Dairy Goat
  Association
Don Wilson, Secretary
P.O. Box 865
Spindale, North Carolina
  28160

American Goat Society, Inc.
J. Willet Taylor,
  Secretary-Treasurer
1606 Colorado
Manhattan, Kansas 66502

PUBLICATIONS

The Dairy Goat Journal
P.O. Box 1908
Scottsdale, Arizona 85252

Better Goat Keeping
Harvard, Massachusetts
  01451

ORGANIZATIONS

Arizona
Arizona State Dairy Goat
  Association
Elaine Hodge, Secretary
2548 W. Hayward Ave.
Phoenix, Arizona 85021

California
Calif. State Dairy Goat Council
Mrs. Alice G. Hall, Secretary
Star Rt. 92334 Box 5B
San Bernardino, California
  92403

Calif. LaMancha Club
Rt. 1, Box 464
Alpine, California 92001

Cenral Coast Co. Dairy Goat
  Association
Roduska Rosales, Secretary
1269 Castroville Blvd.
Salinas, California 93901

Northern Calif. Dairy Goat
  Association
Peggy Blakney, Secretary
Rt. 1, Box 1102
Auburn, California 95603

Pacific Coast Dairy Goat
  Association
Janice Gebhardt, Secretary
752 Corsicana
Oxnard, California 93030

Redwood Empire Dairy Goat
  Association
P.O. Box 6414
Santa Rosa, California 95406

Santa Clara Co. Dairy Goat
  Association
Iris Gillette, Secretary
12895 Columbet Ave.
San Martin, California 95046

Colorado
Western Colorado Dairy Goat
  Association
Toni Hoffman, Secretary
3757½ G Rd
Palisade, Colorado 81526

Connecticut
Connecticut Dairy Goat
  Association
Nancy Tracy, Secretary
RD 2
New Milford, Connecticut
  06776

Illinois
Illinois Dairy Goat Assoc.
Dan Bullock, Secretary
Rt. 1 Box 43
Libertyville, Illinois 60048

Indiana
Indiana Dairy Goat Assoc.
Myrt Wattles, Secretary
Box 21
Ambia, Indiana 47917

TennKylana Dairy Goat
  Association (N. KY. &
  S. Ind.)
Sandy Stoss, Secretary
Rt. 6
Cynthiana, Kentucky 41031

Iowa
Iowa Dairy Goat Association
Norma Zimmerman, Secretary
Rt. 1
Monticello, Iowa 52310

Midwest Dairy Goat Show
  & Producers Inc.
Mrs. Marie Gruber, Secretary
916 Franklin St.
Cedar Falls, Iowa 50613

Kansas
Kansas Dairy Goat Council
Jeff Cross, President
Rt. 1
Riley, Kansas 66531

Maryland
Maryland Dairy Goat
  Association
Gerald Sargent, Jr., Secretary
15711 New Hampshire Ave.
Silver Spring, Maryland
  20904

Massachusetts
Massachusetts Council of Milk
  Goat Breeders' Association
Mrs. Theckla Snell, Secretary
68 Davis St.
Taunton, Massachusetts 02870

Middlesex Co. Dairy Goat
  Breeders Association
Francis Smith, Secretary
49 N. Hancock St.
Lexington, Massachusetts
  02173

Michigan
Michigan Dairy Goat Society
C. Ray Newton, Secretary
1924 Kelsey Highway
Ionia, Michigan 48846

Minnesota
Minnesota Dairy Goat
  Association
Dorothy Jamison, Secretary

Rt. 1, Box 29
Sandston, Minnesota 55072

Nebraska
Mo. River Valley Goat
  Association
Pat Gehrman
Rt. 1
Mead, Nebraska 68041

Nevada
Nevada State Dairy Goat
  Association
Lee Ann Henry, Secretary
20865 Ames Lane
Reno, Nevada 89502

New Jersey
Garden State Dairy Goat
  Association
Audrey Reed, Secretary
715 White Bridge Rd.
Millington, New Jersey 07946

New Mexico
Southwest Dairy Goat
  Association
Robert Jay Wall, Treasurer
P.O. Box 225
Las Cruces, New Mexico
  88001

New York
New York State Dairy Goat
  Breeders Association
Miss Mary Pat Hart, Secretary
456 W. Sand Lake Rd.
Troy, New York 12180

Central New York Dairy
   Goat Society
Mrs. Carol Reardon,
   Secretary
4516 Stony Brook Rd.
RD 2, Oneida, New York
   13421

Ohio
Ohio Dairy Goat Association
Miss Linda McDonnell
Rt. 1, Wheeler Rd., Box 192
LaGrange, Ohio 44050

Firelands Dairy Goat Club
E. R. Schamber
Rt. 6, Bowers Rd.
Mansfield, Ohio 44903

Lorain Co. Dairy Goat Club
Miss Jody Greene
RD 1, Box 197
Wellington, Ohio 44090

Northeast Ohio Dairy Goat
   Club
Mrs. Jean Alspach
3236 Sandy Lake Rd.
Ravenna, Ohio 44266

Northern Ohio Dairy Goat
   Association
Mrs. Raymond H. Klein
10704 Snowville Rd.
Brecksville, Ohio 44141

Scioto Valley Dairy Goat
   Association

Mary Anthony, Secretary
Rt. 3
Marysville, Ohio 43040

Southwestern Ohio Dairy
   Goat Club
Miss Ginger Jackson, President
6984 Thompson Rd.
Cincinnati, Ohio 45247

Western Reserve Dairy Goat
   Club
Mrs. Thelma Calvin
Rt. 2
14251 Clareidon-Troy Rd.
Burton, Ohio 44021

Oklahoma
Oil Capital Dairy Goat Assoc.
D. L. McDonald, President
Rt. 1, Box 209-A
Skiatook, Oklahoma 74070

Oregon
Emerald Dairy Goat
   Association
Dan McMillen, President
Star Rt., Box 197
Dexter, Oregon 97431

Rhode Island
Rhode Island Dairy Goat
   Association
Mrs. Shirley Burnam
Rt. 1, Box 243
Saunderstown, Rhode Island
   02874

Pennsylvania
Pennsylvania Dairy Goat
   Association
Mrs. Edward Watson,
   Secretary
1819 W. Strasburg Rd.
West Chester, Pennsylvania
   19380

Texas
Central Texas Dairy Goat
   Association
Rt. 2
Dublin, Texas 76446

Texas State Dairy Goat
   Association
Mrs. Judy Stubblefield
6809 Green Haven
Amarillo, Texas 79110

Utah
Intermountain Dairy Goat
   Association
Barbara Wilde
P.O. Box 294
Logan, Utah 84321

Vermont
South Vermont Dairy Goat
   Association
Helen Staver
RFD 4
Mtn. Hearth, Malboro
W. Brattleboro, Vermont
   05301

Virginia
Capital Dairy Goat Coop.,
   Inc.
Alexander F. Muir
Murihill Farm
Hillsboro, Virginia 22132

Washington
Inland Empire Dairy Goat
   Association
Pat Hollister, Secretary
Rt. 3, Box 122C
Deer Park, Washington
   99006

Wisconsin
Home and Hobby Dairy
   Goat Club
Dawn Sworski
Rt. 1
Allenton, Wisconsin 53002

INDEX

Abortion, 107
Advanced Register Production
  Record, 24
Afterbirth, 88
Agalactia, 105
American Dairy Goat Associ-
  ation, 11, 83, 94
American Goat Society, 83,
  94
Amino acids, 7
Anemia, 105
Angora goat, 16
Artificial insemination, 83

Bathing, 47
Bedding, 29–31, 67
Bee stings, 106
Bloat, 106
Bran, 85, 108
Breeding, 80–85
  Age, 80
  Linebreeding, 82
  Season, 80, 83, 84, 85
Breeds, 16–23
  Angora, 16
  French Alpine, 16, 18
  La Mancha, 20
  Nubian, 16, 20
  Rock Alpine, 19
  Saanen, 16, 22
  Swiss Alpine, 19 .
  Toggenburg, 16, 25
Bronchitis, 107
Brucellosis, 107
Buck, 81, 96
Bursitis, 107

Butter, 75

Calcium, 7, 45, 85, 113
Canned milk, 79
Castration, 97
Chapped teats, 67
Cheese, 76–79
Chevon, 97
Clipping, 51, 52
Clostridium Perfringens Type
  D, 84, 109
Coat colors, 18, 19, 20, 22
Coccidiosis, 108
Colostrum, 86, 89
Concentrates, 41, 85, 108, 110
Constipation, 43, 44, 85, 108

Daily grooming, 47
Dairy characteristics, 9, 26
Dairy goat show, 13
Dehorning, 53–54, 95
Diarrhea, 92, 108
"Dry-up," 57, 86

Encephalitis, 109
Enterotoxemia, 84, 96, 109
Exercise, 86
Experimental grade, 24
External parasites, 100–102
  Lice, 101
  Mites, 101
  Ringworm, 101
  Screwworms, 102

False pregnancy, 109
Feeder, 34

Fencing, 31–32
Fertilizer, 29
Fluorine Poisoning, 110
Foot Rot, 110
Founder, 45, 102, 110
French Alpine, 16, 18
Frog, 49
Fruits, 43

Gestation table, 84
Goat pox, 111
Goiter, 111
"Gopher" ears, 20
Grade goat, 12, 13
    Experimental, 12
    Half-grade, 12
    Recorded, 12
    Un-recorded, 12
"Graded-up," 24
Grain, 41, 85, 86, 108, 110
Grasses, 39, 45, 46
Grazing, 36, 46
Grooming, 47–53
    Bathing, 47
    Clipping, 51–52
    Daily, 47

Half-grade, 24
Hay, 39–41
    Grasses, 39
    Legumes, 39
Hoof trimming, 48

Importation to America, 8,
    18, 20, 22, 23
    French Alpine, 18
    Nubian, 20
    Saanen, 22
    Tog, 23
Inoculation, 84, 96
Internal Parasites, 98–100
    Blood sucking worm, 99
    Lungworm, 99
    Tapeworm, 99

Ketosis, 112
Kidding, 87–88
    Afterbirth, 88, 110
    Multiple births, 87
    Problems, 88, 111, 112, 113
    Procedures, 88
Kids, 89–96
    Buck, 96
    Castration, 97
    Dehorning, 95
    Feeding schedule, 91
    Formula, 92
    Inoculation, 96

La Mancha, 20
Legume, 39
"Let-down," 57
Linebreeding, 82

Manure tea, 37
Mastitis, 112
Mating, 83
Medication, 104
Milk, 56, 58–60, 65–67, 79,
    105
    Canned, 79
    Equipment, 58, 59, 65
    Filters, 60
    Frozen, 79
    How to, 56
    Off-flavor, 67
    Pasteurizing, 65–66
    Processing, 60–66
    Stand, 59
    Storage, 60–64
    Suppression of, 105
    Weight, 79
Milk fever, 113
Milk production, 6, 10, 16,
    18–19, 21–22, 24, 58
    Daily, 6
    French Alpine, 18
    Nubian, 21
    Peak, 10

Record, 16
Rock Alpine, 19
Sannen, 22
Swiss Alpine, 19
Testing, 24
Yearly, 6
Minerals, 45, 85, 105

Nubian, 16, 20
Nutritional deficiencies, 38, 46, 105

Openness, 9
Osteomalacia, 113

Pasteurization, 65–66
Pasture, 36
Pelleted feeds, 42
Permanent Champion, 24
Photosensitization, 113
Pneumonia, 114
Poisonous plants, 46, 116
"Precocious milker," 80, 86
Pregnancy, 46, 84–87, 107, 109, 114
  Abortion, 107
  Disease, 114
  False, 109
  Inoculation before, 84
Protein, 39, 40, 41, 85
Pulse rate, 104
Purebred, 10, 13

Recorded grade, 12, 24
Registration, 11, 25
Respiration, 104
Rock Alpine, 19
Roughage, 39, 44, 85, 86, 108
Rumen, 39

Saanen, 16, 22
Salt, 34, 45, 114
Salt deficiency, 114
Scours, 92, 108

Scrub goat, 25, 26
Season, 80, 83, 84, 85
Shelter, 27–28
Star Milker certificate, 24
Stripping, 58
Stud fee, 81
Swiss Alpine, 19

Tattooing, 11, 94
Teeth problems, 116
Temperature, 103
Tetanus, 84, 96, 114
Tethering, 36
Toggenburg, 16, 23
Tuberculosis, 115

Udder, 9, 10, 26, 56–58, 67, 81, 106, 112
  Attachment, 10
  Break-down, 26
  Cleaning, 58
  Development, 10, 81
  Dry-up, 57, 86
  Injury to, 56, 106, 112
  Massaging, 57, 58, 67, 112, 115
  Shape, 9
Un-recorded grade, 12

Vegetables, 43–44
Vitamins, 39, 40, 105

Warts, 116
Water, 34, 45
Wattles, 55, 96
Weaning, 95
Weather, 97
Worms, 98–100
  Medication, 100
  Symptoms, 98
  Types, 99